Number Sense Screener™ (NSS™) User's Guide, K-1, Research Edition

Number Sense Screener™ (NSS™) User's Guide, K–1, Research Edition

by

Nancy C. Jordan, Ed.D.
University of Delaware

and

Joseph J. Glutting, Ph.D.
University of Delaware

with

Nancy Dyson, Ph.D.
University of Delaware

·P·A·U·L·H·
BROOKES
PUBLISHING Cº ®

Baltimore • London • Sydney

Paul H. Brookes Publishing Co.
Post Office Box 10624
Baltimore, Maryland 21285-0624
USA

www.brookespublishing.com

::nss˙

Typeset by Integrated Publishing Solutions, Grand Rapids, Michigan.
Manufactured in the United States of America by
H&N Printing & Graphics, Inc., Timonium, Maryland.

For further information about the *Number Sense Screener™ (NSS™), K–1, Research Edition,* please contact Brookes Publishing Co. at www.brookespublishing.com or 1-800-638-3775.

Library of Congress Cataloging-in-Publication Data

Jordan, Nancy C.
  Number sense screener (NSS) user's guide, K–1 / by Nancy C. Jordan and Joseph J. Glutting ; with Nancy Dyson. — Research ed.
     p.   cm.
  Includes bibliographical references and index.
  ISBN 978-1-59857-202-5 — ISBN 1-59857-202-4
  1. Numeracy—Study and teaching (Early childhood)   2. Fractions—Study and teaching (Early childhood)
3. Mathematical ability—Testing.   I. Glutting, Joseph.   II. Dyson, Nancy.   III. Title.
  QA135.6.J67 2012
  372.7—dc23   2012008002

British Library Cataloguing in Publication data are available from the British Library.

2016   2015   2014   2013   2012

10   9   8   7   6   5   4   3   2   1

Contents

About the Authors

Nancy C. Jordan, Ed.D., Professor, School of Education, University of Delaware, 206E Willard Hall, Newark, DE 19716

Nancy C. Jordan is Principal Investigator of the Number Sense Intervention Project (funded by the *Eunice Kennedy Shriver* National Institute of Child Health and Human Development) as well as the Center for Improving Learning of Fractions (funded by the Institute of Educational Sciences). She is author or coauthor of many articles in mathematics learning difficulties and has recently published articles in *Child Development, Journal of Learning Disabilities, Developmental Science, Developmental Psychology,* and *Journal of Educational Psychology.* Dr. Jordan holds a bachelor's degree from the University of Iowa, where she was awarded Phi Beta Kappa, and a master's degree from Northwestern University. She received her doctoral degree in education from Harvard University and completed a postdoctoral fellowship at the University of Chicago. Before beginning her doctoral studies, she taught elementary school children with special needs. Dr. Jordan served on the Committee on Early Childhood Mathematics of the National Research Council of the National Academies.

Joseph J. Glutting, Ph.D., Professor, School of Education, University of Delaware, 221E Willard Hall, Newark, DE 19716

Joseph J. Glutting is a quantitative psychologist. He is a former project director of clinical and industrial measurement for The Psychological Corporation. He is also a certified school psychologist with 5 years' full-time experience in the public schools. He previously taught classes in child psychopathology, intelligence testing, and child personality assessment. Dr. Glutting specializes in applied multivariate statistics and test construction. He developed several standardized measures of intelligence, occupational interest, and attention-deficit/hyperactivity disorder (ADHD) including the *Wide Range Intelligence Test* (*WRIT;* Wide Range, 2000), *Wide Range Interest and Occupation Test–Second Edition* (*WRIOT2;* Wide Range, 2003), and *College ADHD Response Evaluation* (*CARE;* Wide Range, 2002). He coauthored the *Number Sense Battery* (*NSB;* Merrill Publishing, in press) with Nancy Jordan and published more than 100 peer-reviewed journal articles and book chapters. Dr. Glutting currently teaches graduate classes in applied multivariate and univariate statistics, as well as an undergraduate class in tests and measurement. His research is supported by the Institutes of Education Sciences and the National Institutes of Health.

Nancy Dyson, Ph.D., Postdoctoral Researcher, School of Education, University of Delaware, 130 Willard Hall, Newark, DE 19716

Nancy Dyson has been in education for more than 30 years as both a teacher and the director of a parent cooperative school. She recently completed her doctoral degree in education at the University of Delaware with a research focus on students struggling with mathematics.

Acknowledgments

The development of the Number Sense Screener™ (NSS™) was supported by grants from the *Eunice Kennedy Shriver* National Institute of Child Health and Human Development (Grants HD036672 and HD059170). We thank our participating children and schools for their extraordinarily generous cooperation.We also thank Maria Locuniak and the University of Delaware graduate assistants who assisted us with data collection and analysis.

CHAPTER 1

Introduction to the Number Sense Screener™ ∷nss™

The Number Sense Screener™ (NSS™) is a research-based tool for screening early numerical competencies in kindergarten and early first grade. These competencies predict achievement level and growth in elementary school mathematics. The areas assessed on the NSS are aligned with the Common Core State Standards for kindergarten mathematics (National Governors Association Center for Best Practices & Council of Chief State School Officers, 2010) and thus are uniquely suited for planning interventions and measuring progress. The NSS is a standardized measure designed to be used by teachers, school psychologists, learning specialists, and other school-related personnel.

The NSS includes 29 items with norms for the fall and spring of kindergarten and the fall of first grade. It is derived from a longer research instrument. Rasch item analyses, as well as a more subjective review of issues related to item bias, were used to choose the final items for NSS (Jordan, Glutting, & Ramineni, 2008). It is individually administered and takes 15–20 minutes to conduct.

The Math Problem

Underachievement in mathematics can have serious educational and vocational consequences. Students with weak mathematics skills at the end of middle school are less likely to graduate from college than are students who are strong in mathematics (National Mathematics Advisory Panel, 2008). Proficiency in advanced mathematics is important for success in college-level science courses and a wide range of vocations in the sciences (Sadler & Tai, 2007). Many students in U.S. elementary schools do not develop adequate foundations for success in mathematics, and low-income learners lag far behind their middle-income peers (National Center for Educational Statistics, 2009).

Fluency with mathematics operations is a hallmark of mathematical learning in the early grades. Lack of fluency is a signature characteristic of students with learning disabilities in math (Jordan, 2007). Research suggests that calculation deficiencies, from the first year of formal schooling onward, can be traced to fundamental weaknesses in number sense (Gersten, Jordan, & Flojo, 2005; Malofeeva, Day, Saco, Young, & Ciancio, 2004). Weak number sense results in poorly developed counting procedures, slow fact retrieval, and inaccurate computation (Geary, Hamson, & Hoard, 2000; Jordan, Hanich, & Kaplan, 2003a, 2003b). It is difficult to memorize arithmetic "facts" by rote without understanding how combinations relate to one another on a mental number line. Accurate and efficient counting procedures can lead to strong connections between a problem and its solution (Siegler & Shrager, 1984). It has been suggested that basic number sense is a circumscribed cognitive function, one that is relatively independent from general memory, language, and spatial knowledge (Gelman & Butterworth, 2005;

Landerl, Bevan, & Butterworth, 2004). Although there is a high rate of co-occurrence between math and reading or language difficulties, specific math difficulties with typical development in other cognitive and academic areas are well documented (Butterworth & Reigosa, 2007; Jordan, 2007).

Definition of Number Sense

Although *number sense* has been defined differently in the research literature (Gersten et al., 2005), it is generally agreed that number sense in the 3- to 6-year-old period involves inter-related abilities involving numbers and operations, including recognition of small quantities, counting items in a collection to at least 5 with knowledge that the final count word indicates how many are in the set, discriminating between small quantities (e.g., 4 is greater than 3, 2 is less than 5), comparing numerical magnitudes (e.g., 5 is 2 more than 3), and transforming sets of 5 or less by adding or taking away items. In its 2009 report, *Mathematics Learning in Early Childhood: Paths Toward Excellence and Equity* (National Research Council, 2009), the Committee on Early Childhood Mathematics underscored the importance of number sense for success to school mathematics. Weak number sense prevents children from benefiting from formal instruction in mathematics (Baroody & Rosu, 2006; Griffin, Case, & Siegler, 1994).

Most children bring considerable number sense to school, although there are individual differences associated with social class and learning abilities (e.g., Ginsburg & Golbeck, 2004; Ginsburg & Russell, 1981; Jordan, Huttenlocher, & Levine, 1992; Jordan & Levine, 2009). Non-verbal number sense appears to be present even in infancy (Mix, Huttenlocher, & Levine, 2002). Preverbal infants can discriminate between two and three objects and are sensitive to ordinal relations and number operations (Starkey & Cooper, 1980; Wynn, 1992). Knowledge of the verbal or symbolic number system, however, is heavily influenced by experience or instruction (Geary, 1995; Levine, Jordan, & Huttenlocher, 1992). Symbolic number sense development is closely associated with children's home experiences with number concepts (Case & Griffin, 1990), and efforts to teach number sense to high-risk children in early childhood have resulted in significant gains on first-grade math outcomes compared to control groups (Dyson, Jordan, & Glutting, in press; Griffin et al., 1994).

Elements of Number Sense

Counting is important to early mathematics because it extends quantitative understanding beyond very small numbers (i.e., 1, 2, 3), which initially are recognized as whole units (Baroody, 1987; Baroody, Lai, & Mix, 2006; Ginsburg, 1989). Before kindergarten, most children grasp that each object in a set is counted once and only once, that the count words are always used in the same sequence (i.e., 1, 2, 3, and so forth), and that the last number in the count always denotes the number of objects in the set (Gelman & Gallistel, 1978). Children gradually learn that they can count objects presented in any configuration as long as they count each object only once.

Being able to compare the magnitudes of quantities and numerals is also important for learning mathematics (Case & Griffin, 1990; Griffin, 2002, 2004). Many young children can tell which of two piles of objects has more or less and that the written number 5 is bigger than 3 or that 3 is smaller than 5 (Griffin, 2004). Linking quantities, number words, and numerals helps children see relations between and among numbers on the number line. Children learn that the next number in the count sequence or on a number list is always one more than the preceding number (i.e., they form a linear representation of number). This knowledge helps children use counting to solve addition and subtraction ($n + 1$; $n - 1$) and learn answers to combinations based on relations between numbers in the count sequence (Baroody, Eiland, & Thompson, 2009).

Competence with numerical operations is of primary importance to mathematics in elementary school. Addition and subtraction abilities start to develop in early childhood. Although preschoolers have little success with verbally presented story problems (e.g., "Bob had 2 pennies. Jenny gave him 1 more penny. How many pennies does Bob have now?") and number combinations (e.g., "How much is 2 and 1?"), many can solve nonverbal calculations with object representations (e.g., the child is shown two dots [tokens] that are then hidden with a cover, and when one more dot is slid under the cover, the child determines how many objects are now under the cover; Levine et al., 1992). Even children with limited counting facility can solve these kinds of problems with totals of 5 or less, most likely by visualizing sets when objects have been added or taken away (Huttenlocher, Jordan, & Levine, 1994).

Success in mathematics requires children to connect understandings of object-based calculations to verbal story problems and number combinations. Low-income children as well as those with specific language impairments typically perform below their middle-income peers on story problems and number combinations (Jordan, Kaplan, Olah, & Locuniak, 2006; Jordan, Levine, & Huttenlocher, 1994; Starkey, Klein, & Wakeley, 2004) but not on nonverbal problems (Jordan et al., 1992). High-risk learners do not make use of counting strategies to help them calculate with larger numbers (Jordan et al., 2006). Early intervention should help children link their developing knowledge of addition and subtraction, as demonstrated on the nonverbal problems, to conventional verbal combinations through counting and mental representation.

Predictability of Number Sense

Number sense performance and growth in kindergarten and first grade is highly predictive of mathematics achievement through at least third grade, even when adjusting for reading, age, and general cognitive factors (Jordan, Glutting, & Ramineni, 2009; Jordan, Kaplan, Ramineni, & Locuniak 2009; Locuniak & Jordan, 2008). Recent work suggests that the connection may last through adulthood (Duncan et al., 2008). Moreover, number sense in preschool predicts performance on similar measures in kindergarten (VanDerHeyden, Broussard, & Cooley, 2006). The majority of children lacking in number sense are from low-income populations (Jordan, Kaplan, Locuniak, & Ramineni, 2007; Jordan et al., 2006). Kindergarten number sense is critical for setting children's developmental learning trajectories in elementary school mathematics. Early weaknesses can trigger more pervasive problems as development progresses (Thomas & Karmiloff-Smith, 2003). Children who leave kindergarten with low number sense may enter first grade at a disadvantage and never catch up to children who started with adequate number sense.

Response to Intervention

The NSS is relevant to response to intervention (RTI) service delivery models. The Individuals with Disabilities Education Improvement Act (IDEA) of 2004 (PL 108-446) encourages schools to use RTI for identifying students with learning disabilities as an alternative to models emphasizing discrepancies between a student's IQ and achievement. The NSS can help professionals to 1) screen and identify students at risk for math difficulties early, 2) develop evidence-based interventions (see Dyson et al., in press), 3) monitor progress at regular intervals from the fall of kindergarten to the fall of first grade, and 4) adjust instruction according to students' responsiveness. (See Chapter 3 regarding how to use NSS findings to develop interventions.)

Test Content and Organization

The NSS is organized around the following number topics: Counting Skills, Number Recognition, Number Comparisons, Nonverbal Calculation, Story Problems, and Number Combinations.

Counting Skills

Counting Skills involves three simple items. First, children are shown a stimulus page with five stars printed on it. They are asked to count each star and to touch each star as they count. Then the examiner turns to a blank page and asks the child, *"How many stars were on the paper you just saw?"* The task determines whether children can enumerate or count a collection and know the *cardinality principle*. Children are next asked to count as high as they can but are stopped at 10. Children are allowed to restart counting only once but are always allowed to self-correct. For more diagnostic information and progress monitoring, the examiner can also ask children to count higher, but only knowledge of the sequence from 1 to 10 is scored.

Number Recognition

Number Recognition requires children to name 13, 37, 82, and 124. The child is given one point per number. The examiner may start with smaller numbers such as 2, 4, and 9 to allow the child to start with more familiar numbers. However, credit is given only for naming the four numbers listed in the protocol.

Number Comparisons

Number Comparisons, adapted from Griffin (2002), consists of seven items. Shown the number 7, children are asked what number comes right after 7 and what number comes two numbers after 7. Shown two numbers (5 or 4; 7 or 9), children are asked which number is bigger. They also are asked which of two numbers is smaller (8 or 6; 5 or 7). Finally, children are shown a visual array of three numbers spaced equidistantly (5, 6, and 2) and asked which number is closer to 5: 6 or 2.

Nonverbal Calculation

At the start of *Nonverbal Calculation*, the examiner places a white mat in front of the child and opens a small box of 10 black "dots"(tokens), placing it off to the side. The cover of the box has an opening cut into the side through which dots can be pushed into the box. Instructions for making these materials can be found in Appendix A of this *Number Sense Screener™ (NSS™) User's Guide, K–1, Research Edition.*

The *Number Sense Screener™ (NSS™) Stimulus Book, K–1, Research Edition* contains pages with arrays of horizontal dots. The examiner begins by telling the child that they will be playing a game with the dots. On the demonstration item, the examiner places three dots on the mat in a horizontal row. The dots are then hidden with a cover. The child is asked to point on a stimulus page to how many dots are hiding under the cover. The stimulus page has four choices, each enclosed in a rectangle. The child is corrected on this demonstration item if the wrong answer is given, to make sure he or she understands the multiple-choice format. Three addition problems and one subtraction problem are then presented (2 + 1, 3 + 2, 4 + 3, 3 − 1). The examiner places a number of dots on the mat (in a horizontal line) and tells the child how many dots are on the mat. The examiner then covers the dots with the box cover. The examiner places a second group of dots (also in a horizontal line) on the mat near the opening in the box cover and tells how many dots he or she put down. The examiner then says, *"Watch what I do!"* and pushes the dots through the cover's side opening one by one. For the subtraction problem, the examiner says, *"Watch what I do!"* and then pulls out a dot through the opening and says, *"One dot."* For each item, the child is asked to indicate how many dots are hiding under the box cover, either by pointing on the NSS stimulus book to the collection with the same number of dots or by stating the number word. Because this is a nonverbal task, the child is not penalized if he or she points to the correct rectangle with dots but says the wrong word. The nonverbal calculation task was adapted from Levine et al. (1992).

Story Problems

Story Problems includes three oral addition (2 + 1, 4 + 3, and 3 + 2) and two oral subtraction (6 – 4 and 5 – 2) story problems, read one at a time. Children are told they can use their fingers, a number list, or a pencil and paper to help them find the answer. The addition problems are phrased simply, following this basic format: "*Jill has **m** pennies. Jim gives her **n** more pennies. How many pennies does Jill have now?*" Similarly, the subtraction problems are phrased, "*Kisha has **m** pennies. Peter takes away **n** of her pennies. How many pennies does Kisha have now?*"

Number Combinations

In the *Number Combinations* section, four addition (2 + 1, 3 + 2, 4 + 3, and 2 + 4) and two subtraction (7 – 3 and 6 – 4) number combinations are orally phrased as "*How much is **m** and **n**?*" and "*How much is **n** take away **m**?*" As with Story Problems, children are told they can use their fingers, a number list, or a pencil and paper to help them find the answer.

Although the child is given credit only for correct responses on Story Problems and Number Combinations, the examiner is encouraged to write down the strategy that the child uses spontaneously. Finger use in kindergarten is highly adaptive and predicts mastery of addition and subtraction in first and second grade (Jordan et al., 2008). Other adaptive strategies for kindergartners and first graders include counting on from an addend (e.g., for 3 + 2, saying 3 and then counting 4, 5), use of a number list for counting, and using paper and pencil to write down tally marks or circles (Dyson et al., in press).

Kindergarten Common Core State Standards

The content of the NSS is closely aligned to the Common Core State Standards for Mathematics Kindergarten: Counting and Cardinality, Operations and Algebraic Thinking, and Number and Operations (National Governors Association Center for Best Practices & Council of Chief State School Officers, 2010). According to the Common Core, kindergartners are expected to know number names (NSS Number Recognition subarea), to know the count sequence and be able to count to tell the number of objects in a collection (NSS Counting Skills subarea), and to compare numbers presented as written numerals between 1 and 10 (NSS Number Comparisons subarea). In terms of operations, kindergartners are expected to "understand addition as putting together and adding to and understand subtraction as taking apart and taking from" (National Governors Association Center for Best Practices & Council of Chief State School Officers, Operations and Algebraic Thinking section; NSS Nonverbal Calculation, Story Problems, and Number Combinations subareas). Kindergartners also are expected to work with numbers between 11 and 19 to gain foundations for place value (NSS Number Recognition subarea).

National Council of Teachers of Mathematics Focal Points

The NSS aligns with the Kindergarten Focal Points of the National Council of Teachers of Mathematics (NCTM) in the area of numbers and operations. This includes representing, comparing, and ordering whole numbers (NSS Counting Skills, Number Recognition, and Number Comparisons subareas) and joining and separating sets (NSS Nonverbal Calculation, Story Problems, and Number Combinations subareas).

User Qualifications

A wide range of providers—including learning specialists, classroom teachers, and school psychologists—can use the NSS. Users of the NSS should have training in assessment and interpretation of test results. When the NSS is employed for screening purposes, it is important that interpreters share results in terms of "risk factors" for mathematics difficulties and not overly interpret findings to provide a diagnosis. When the NSS is employed as part of a comprehensive evaluation, user qualifications are guided by the certification and licensing requirements of the evaluator's discipline.

Standardization

Norms for the NSS are unusual in that longitudinal (versus cross-section) sampling was employed. Norms were obtained in three stages, wherein the same children were evaluated on each occasion: Fall 2003 ($N = 367$), Spring 2004 ($N = 435$) and Fall 2004 ($N = 300$). The standardization model accounted for age, gender, race or ethnicity, and region of residence. Demography detailed in the following represents children participating in the final test phase, where $N = 300$.

Age: All children were evaluated in the fall of their kindergarten year, again in the spring of kindergarten, and a third time in the fall of first grade. The mean ages (years-months) at each time frame were, respectively, 5-8 ($SD = 3.7$), 6-2 ($SD = 3.7$), and 6-8 ($SD = 3.7$).

Gender: The overall standardization sample included 135 females (45.0%) and 165 males.

Geographic Region: All students taking part in the standardization effort came from the state of Delaware.

Race or Ethnicity: Four categories of race or ethnicity were monitored. The sample was 42.6% Anglo-American, 36.8% African American, 15.5% Hispanic, and 5.1% Asian.

Socioeconomic Status: Children's school lunch program (free or reduced lunch versus full pay) was used a proxy for socioeconomic status (SES). Of the entire sample, 42.2% ($n = 124$) received free or reduced lunches. Evident here is that the NSS's sample is heavily weighted to children with lower SES backgrounds.

English Language Learners: There were 22 children in the sample who were classified by their schools as having limited English proficiency. This number corresponds to 7.3% of the standardization sample.

Classification Status: Of students in the NSS's standardization sample, 12.9% ($n = 38$) were receiving special education services.

Standard Scores

Standard scores for the NSS were normalized to the normal (bell-shaped) distribution in the population. Specifically, a cumulative frequency distribution of raw scores was produced for each of the three age groups in the standardization sample. Next, each variable was transformed so that its shape matched the bell-shaped curve. Finally, norms tables were developed by comparing each standard score to its corresponding raw score. This process of transforming raw scores to normalized standard scores represents the most common application of a "nonlinear area conversion" (Thorndike, 1982, p. 115). In addition to changing the first two moments of the raw-score distribution (M and SD), normalized area conversions also adjust the amount of skewness and kurtosis. Specifically, skewness is adjusted to the normal (bell-shaped) curve value of 0 rather than to a positively skewed value ≥ 1.00 or to a negatively skewed value ≤ 1.00. Likewise, kurtosis is set to the normal-curve value of 0 rather than to a leptokurtotic value of ≥ 1.00 or to a platykurtotic value ≤ 1.00.

The NSS provides both standard scores and percentiles. A principal advantage of the normalized transformations used with the NSS is that percentiles corresponding to identical standard scores are equal, because they follow well-known properties of the bell-shaped curve. Thus, all normalized standard scores of 115 will hover around a percentile rank of 68, and all normalized standard scores of 130 will hover around a percentile rank of 98.

CHAPTER 2

Administration and Scoring Procedures

A professional who is thoroughly familiar with the administration procedures gives the NSS™ individually. Kindergarten children are usually happy with the personal attention they receive through participating in this assessment. This positive experience depends on the examiner keeping an upbeat demeanor and encouraging the child to do his or her best. Children should never get a sense from the examiner that they are performing below standard.

Testing Materials

Testing materials include

- *Number Sense Screener™ (NSS™) Stimulus Book, K–1, Research Edition,* a spiral-bound book containing visual stimuli
- *Number Sense Screener™ (NSS™) Record Sheet, K–1, Research Edition,* a sheet for recording student responses and for scoring results
- *Number Sense Screener™ (NSS™) Quick Script, K–1, Research Edition,* a book with explicit, verbatim instructions for each item and subarea
- A box with 10 black dots (tokens) and a white foam mat (approximately 9 × 12 inches) for the Nonverbal Calculation task; the box cover has an opening in the side for sliding the dots in and out. Instructions for making these materials can be found in Appendix A of this *Number Sense Screener™ (NSS™) User's Guide, K–1, Research Edition.*
- Worksheet and pencil for the Story Problems and Number Combinations subareas; a master copy of the worksheet can be found in Appendix B of this NSS user's guide
- A number list for the Story Problems and Number Combinations subareas; a master for making a number list can be found in Appendix C of this NSS user's guide

General Instructions for Administering the Number Sense Screener™ and Recording Responses

The positioning of the child and the examiner is important, because the child must be able to see the NSS stimulus book and the examiner must be able to see how the child is interacting with the NSS stimulus book. It is easiest for the child to sit around the corner of the table from the examiner. If the examiner is right handed, the child should be seated to the left of the examiner, and vice versa. The table should have plenty of space for the child to have the white mat in front of the NSS stimulus book and room for the examiner to write on the NSS record sheet. The examiner places the NSS stimulus book in front of the child. Pages are turned from top to bottom, not left to right. The examiner should make sure the child's view of the pages is not blocked.

An exact script is provided in the NSS quick script, and it is imperative to follow the protocol for the following reasons: 1) the protocol has been developed through research and has been shown to be the clearest way to present the questions; 2) if the directions in the protocol are not read verbatim, children may be led inadvertently to give correct answers they otherwise might not have given. Children should not be given hints or in any way led to correct answers. Hints include gestures, expressions, or any indication that the child has been successful or unsuccessful. The examiner should not indicate whether the child is right or wrong, but rather give praise for listening or for working hard.

The examiner reads each question slowly and makes sure the child is paying attention. (A question can be repeated once if the child requests it or if it is clear he or she was not attending.) If necessary, children should be reminded throughout the test to listen carefully. Although the test is not timed, the child should be encouraged to respond after about 10 seconds, even if the response is "I don't know." If the child still does not respond, the problem is marked incorrect.

Instructions by Subarea

The examiner begins the assessment by introducing himself or herself and addressing the child by name. Say, *"We are going to play some number games. It is important that you listen carefully and do your best. Are you ready to play?"* As the examiner moves from subarea to subarea, he or she must be sure to follow the NSS quick script as it contains transitional phrases. Also, the examiner must be sure, when indicated, to turn to transitional blank pages in the NSS stimulus book. Children should not be looking at previous material while the examiner is introducing the next subarea.

Counting Skills

For Counting Skills item 1, the examiner says, *"Here are some stars. I want you to count each star. Touch each star as you count."* It is important that the examiner is able to see how the child is touching the stars to be sure he or she is using one-to-one correspondence. In the five blanks to the right of item 1 on the NSS record sheet, write in the numbers the child says as he or she counts. If the child stops before 5, put a dash in the remaining blanks. In the blank under "Correct (+)/Incorrect (#)" in the NSS record sheet, the examiner should write a + if the child performs the count correctly and a – if the child makes any error. For example, if the child counts, "One, two, three, four, five," record the response as follows:

1. Write each number said when child is asked to count the stars _1_ _2_ _3_ _4_ _5_ (remember to record a + in the correct/incorrect box)

However, if the child says, "One, two, four, five, six" fill in the "count" blanks as follows:

2. Write each number said when child is asked to count the stars _1_ _2_ _4_ _5_ _6_ (record a – in the correct/incorrect box)

The examiner then turns to a blank page in the NSS stimulus book to Counting Skills item 2 and asks, *"How many stars were on the paper you just saw?"* The child is looking at a blank page, so he or she may say "none" or "zero." If the child gives such a response, it is important to know whether the child understands that the examiner is referring to the previous page, not the one in front of them. In cases such as this, the examiner should say, *"Not on this paper. How many stars were on the paper you just saw?"* Record a + if the child correctly responds "five." Record the number he or she says if the response is incorrect.

For Counting Skills item 3, the examiner says, *"I want you to count as high as you can. But I bet you're a very good counter, so I'll stop you after you've counted high enough, Okay?"* If the child counts to

10 without an error, record a +. If the child makes an error, record the last correct number spoken. In order to encourage children, let them continue to count to 10 even if they make an error. Then say, *"Thank you; that's high enough."* For example, if a child counts, "one, two, three, four, five, six, eight, nine, ten," it would be recorded as follows:

3. _6_ (10)

For more diagnostic information and progress monitoring, the examiner can also allow children to count higher, but only knowledge of the sequence from 1 to 10 is scored. Suppose a child counts correctly to 10 and is allowed to continue. He or she then counts correctly to 12 but then says "14" instead of "13." This response would be recorded as follows:

Max count (optional) _12_

Number Recognition

The examiner begins by saying to the child, *"I'm going to point to some numbers. I want you to name the number when I point to it. Some may be hard for you, so don't worry if you don't get them all right."* For the four items in the Number Recognition subarea, the examiner shows the child a number in the NSS stimulus book, points to the number, and says, *"What number is this?"* If a child responds correctly, a + is recorded, but if he or she responds incorrectly, his or her exact response is recorded. This may require the examiner to write words and not just numerals, because the child may give a response that can not be captured with numerals (e.g., the child may say "twelve-four" instead of "one hundred twenty-four"; in order to ensure that the child's response is captured accurately, these sorts of responses should be written out in words).

Smaller "practice" numbers (2, 4, and 9) are included to allow the child to begin the task with more familiar numbers and for diagnostic purposes. These numbers are for practice only and, although the child's responses should be recorded, credit is only given for naming the four numbers listed as items 1–4 (13, 37, 82, and 124) in the NSS quick script protocol and on the NSS record sheet.

Number Comparisons

The seven items in the Number Comparisons section assess understanding of number magnitude when presented in a symbolic form. The tasks are adapted from Griffin and Case (2002). Numerals corresponding to the numbers in the tasks are presented in the NSS stimulus book. However, no physical or pictorial model of the quantities is given. The section begins with the examiner saying to the child, *"Now I am going to ask you about some more numbers. Listen carefully and do your best."* During this transition, the child is shown a blank page. As before, if a child responds correctly, a + is recorded, but if he or she responds incorrectly, his or her exact response is recorded.

For item 1, the examiner shows the number 7 and says to the child, *"What number comes right after 7?"* It is important to say the words *right after* and not just *after*, because there are many numbers that come "after" 7, but only one number that comes "right after." For item 2, the examiner continues to show the number "7" and asks, *"What number comes two numbers after 7?"* Some children may respond "8 and 9" or simply "8, 9." Although the correct response is "9," these answers are also counted as correct, because they show that the child knows both of the numbers after 7 in the correct order.

For item 3, children are shown the numbers 5 and 4 and are asked, *"Which is bigger or more, 5 or 4?"* The examiner should point to the numbers in the NSS stimulus book as they are read. The procedure is repeated for item 4 using the numbers 7 and 9. Before moving to item 5, a blank page is shown in the NSS stimulus book while the examiner says, *"This time I'm going to ask you about smaller numbers."* The word *smaller* should be emphasized to help children shift to the new task. Items 5 and 6 are performed in the same manner as items 3 and 4, but now with the examiner asking, *"Which is smaller or less?"*

For item 7, children are presented with the numbers 5, 6, and 2 . Children are asked *"Which number is closer to 5: 6 or 2?"* Some children may try to measure which number is physically closer. If that occurs, the examiner should redirect those children: *"Not closer that way; which is closer to 5 in number?"*

Nonverbal Calculation

Children especially enjoy this part of the assessment, as it has the appearance of a game. Children will be asked to visualize transformations of sets and to select an answer from a set of four possible answers. The answers are represented as a linear array of black dots corresponding to the black dots or tokens that will be used in the task. The child is *not* required to give a verbal response; he or she must simply choose an answer out of a choice of four and point to it. As before, if a child responds correctly, a + is recorded, but if he or she responds incorrectly, his or her exact response is recorded.

The NSS stimulus book is turned to a blank page. The examiner begins by placing a white mat (see the materials list earlier in this chapter) in front of the child. The NSS stimulus book must be placed so that the child can see it and the mat simultaneously. The examiner begins by saying, *"We are going to play a game with these dots. Watch carefully."*

In this subarea, the first item is a practice item to ensure that children understand the multiple-choice format. The examiner places three black dots (tokens) in a horizontal line on the mat and says, *"See this...there are three dots."* The child is permitted to observe the dots for about 3 seconds. Then the examiner covers the dots with the box cover and turns to the next page in the NSS stimulus book. On this page there are four rectangles, each with a horizontal array of 1, 2, 3, or 4 dots. The examiner then says, *"Now point to the number of dots that are in the box."* If a wrong answer is given, the child should be corrected. This should be noted for diagnostic purposes, but the practice item is not included in the subtest score.

For item 1 (2 + 1 = 3), the page in the NSS stimulus book should first be turned so that the child is looking at a blank page. Two dots are placed on the mat in a horizontal line. The examiner says, *"See this...there are two dots."* The child is given about 3 seconds to observe the dots. The examiner then covers the dots with the box cover. One more dot is put down and the examiner says, *"Here is one more dot."* Before sliding the dot under the cover, the examiner says, *"Watch what I do."* The dot is then slid under the box cover in full view of the child.

The blank page in the NSS stimulus book is now turned to reveal the multiple-choice page. The examiner says, *"Point to the number of dots that are hiding in the box now. Look at all the choices before you pick."*

Items 2 (3 + 2 = 5) and 3 (4 + 3 = 7) are performed in the same manner, replacing the underlined words with the appropriate addends for the task. (When there are more than one to add, the dots are slid in one at a time.) The NSS stimulus book page must be turned before each task to reveal another blank page.

Item 4 models subtraction (3 − 1 = 2). In order to help the children shift to a new model, the examiner says, *"Now we are going to do something a little bit different. Watch carefully."* Three dots are placed on the mat in a horizontal line. The examiner says, *"See this? There are three dots."* After about 3 seconds the dots are covered by the box as before. Unlike before, no new dots are put down. Instead, the examiner says, *"Watch what I do."* The examiner takes out one dot from under the box cover in full view of the child and says, *"See, here is one dot."*

Note that, because this is a nonverbal assessment, the child is not penalized if he or she points to the correct box but says the wrong answer. In the same way, if he or she points to the wrong box and says the correct answer, it should be scored as correct. The child may have known the correct answer but was quick to choose a box without carefully counting the dots. Any such inconsistencies should be noted in the margin of the NSS record sheet.

Story Problems

The items in the Story Problems subarea assess the child's ability to pull the mathematics out of a story and perform either an addition or a subtraction problem. The NSS stimulus book must be turned to a blank page before beginning this subarea. Children are given an option of using one or more of three different strategies to help them solve the problem (see the materials list earlier in this chapter). The examiner puts a number list, a piece of paper, and a pencil in front of the child and says,

> *I'm going to give you some story problems. You can use your fingers (holds up fingers), or this number list (gestures to number list), or this paper and pencil to help you find the answer. Some questions might be easy for you and others might be hard. Don't worry if you don't get them all right. Listen carefully to the question before you answer.*

The examiner should not hand the pencil to the child but rather place it on the table. If the child does not want to use the pencil, holding it can be a distraction. Handing the child the pencil also indicates that the examiner expects the child to use it.

Each story problem is read aloud by the examiner at a slow pace. Items may be repeated once if requested or if it is clear the child was not listening. The child must be observed while solving the problem to determine which, if any, tool or strategy the child uses to solve the problem. Children should be given adequate time to solve the problem using the strategy they have chosen. Drawing typically takes considerably more time than counting on fingers or counting on the number list. If the child decides to use the worksheet (see Appendix B), the examiner should encourage the child to begin in the upper left hand box by saying, *"Work in this box."* Most children instinctively begin in the upper left corner. The examiner should allow children to use only one box per problem. If they begin working in a new box while still on the same problem, the examiner should point to the original box and say, *"Stay in this box."* In some cases, children begin acting on the first number mentioned in the problem (e.g. drawing or making the number on their fingers) and they may not hear the second half of the problem. If the child asks to have the second addend repeated by saying things such as, "How many more pencils?" it is appropriate to repeat that part of the problem, but *not* to give him or her just the number. The child must extract the number from the sentence. For example, in the case of item 1, the examiner says, *"Jill has 2 pennies. Jim gives her 1 more penny. How many pennies does Jill have now?"* If the child asks, "How many more pennies?" the appropriate response would be *"Jim gives her 1 more penny"* and *not* simply saying "1." The examiner must also be careful not to reword the problems or give the child any hints, such as stating the combination. For example, for item 1, the examiner may not say "two plus one" or "one more than two." The examiner may not in any way do the job of extracting the numbers from the problem for the child.

As always, if a child responds correctly, a + is recorded, but if he or she responds incorrectly, his or her exact response is recorded. Strategies used by the child to solve the problem should be circled on the NSS record sheet. A more complete discussion of possible strategies is provided in the next section in this chapter, Recording Strategies. The examiner should not suggest to a child to try a strategy, other than giving the opening statement about the tools he or she is providing. For example, if a child seems lost and is staring into space for one of the problems, it should *not* be suggested to the child to count on his or her fingers or to use any other tool.

The examiner reads the story problem and the child gives an oral response after trying to solve the problem. Items 1–3 are addition (join) problems. Items 4 and 5 are subtraction (take away) problems. To alert children of the shift, when moving from item 3 to item 4, the examiner says, *"Now these questions are a little different. Listen carefully."*

Number Combinations

The Number Combinations subarea assesses the child's ability to manipulate sets of small numbers presented in a symbolic form. The same initial statement about strategies as that in the Story Problems subarea is given, and the children keep the same tools. If the child chooses to use the worksheet, the examiner should direct the child to the upper left hand box in the Number Combinations portion of the Story Problems and Numbers Combinations Worksheet (see Appendix B).

The NSS stimulus book has the combinations written in numeric form, and the examiner reads the combinations while pointing to the numbers and plus or minus symbols in the NSS stimulus book. The plus (+) sign is read "and" and the minus (–) sign is read "take away." If the child responds "21" for "two and one" the examiner should clarify by saying, *I am looking for 2 plus 1. How much is 2 and 1?"*

To begin this subarea, the NSS stimulus book should still be on a blank page. The examiner turns the page and says, *"How much is 2 and 1?"* while touching the numbers and the plus sign.

This protocol continues until the addition problems (items 1–4) are complete. Then the examiner turns to a blank page and says, *"Now these questions are a little different. Listen carefully."* Items 5 and 6 are subtraction problems.

As before, if a child responds correctly, a + is recorded, but if he or she responds incorrectly, his or her exact response is recorded. Strategies the child uses to solve the problem should be circled on the NSS record sheet (see the next section, Recording Strategies). Again, the examiner must not suggest the child try a strategy (e.g., finger counting or use of any other manipulative) other than giving the opening statement about the tools available to him or her.

Recording Strategies

Although not considered in the norms, the NSS allows for observation and recording of the child's problem-solving strategies for Story Problems and Number Combinations. Observing the strategies a child uses (and does not use) helps evaluators get a better view into the student's understanding of number and number operations. The coding categories describing young children's calculation strategies are listed on the NSS record sheet to allow the examiner to simply circle the observed strategy. If a strategy is observed that is not listed on the document, the examiner should write the observed strategy in the "Notes" section next to each question on the NSS record sheet.

The following guidelines aid in recording student strategies. There are two categories of strategies: 1) the tool the child is using (written combination, drawing, number list, fingers, or counting without a tool) and 2) the method of counting (quick retrieval, counting—all, counting—on). After the examiner notes the tool the child is using, he or she then should observe what counting method the child is using. Sometimes a child uses more than one tool paired with a counting method. Every strategy used should be circled.

Counting Tools

Written Combination

The student writes on paper the numbers corresponding to a given story problem and separates each number with or without a plus or minus sign. For example, for the story problem *José has 3 cookies. Sarah gives him 2 more cookies. How many cookies does José have now?* the child may write "3 + 2" or simply "3 2." The examiner circles *written combination,* but must then carefully observe how the child obtains the sum. Some children may simply state the sum, "5," which would be recorded as *quick retrieval;* others may count on their fingers (see following section on finger counting). Each strategy used should be recorded. If a child simply copies the combinations during the Number Combination subarea of the test, do not mark this strategy. The child did not need to extract the combination from the problem but merely copied the combination given in the NSS stimulus book. This is not using written combination as a strategy.

Drawing

The student draws figures (e.g., circles, sticks, pencils) to represent quantities in the story problem.

Number List

The student physically uses the number list provided to answer the given problem. For example, for a story problem that corresponds to 3 + 2, the child may touch the 3 or the dot under 3 or count up to 3 touching three dots. Then the child counts up two more, landing on 5. Touching the number list, however, does not always mean that the child is using it as a problem-solving tool. A child may touch the 3, then touch the 2, then give an answer. Touching the number list in this way is not using it as a tool to solve the problem, and in this case *number list* should *not* be circled.

Fingers

The student uses his or her fingers to aid in counting.

Counting without a Tool

The student counts out loud without touching a tool. This may also be circled if the child is obviously counting silently. For example, the child may bob his or her head or tap the table.

Counting Methods

Quick Retrieval

The student gives a correct answer quickly without using any type of strategy. The child may also say something like, "I know this one," and give a quick answer.

Counting—All

The student creates both addends either by drawing or on fingers, then counts all the objects beginning with "one." For example, the student makes 3 on one hand and 2 on the other hand and then counts 1, 2…3, 4, 5 on fingers to arrive at the answer, 5.

Counting—On

The student says and/or makes the first addend on his or her fingers and starts counting upward from the first addend to determine the sum. For example, for 3 + 2, the student makes 3 on his or her fingers, then says "3…4, 5," or the student simply says "3" and then counts up using fingers "4, 5." A student may also draw three circles and then say "3…4, 5." Students may also use a counting—on strategy to solve subtraction problems. For example, if the problem was 6 – 4, the child would start with 4, then count up until he or she reaches 6, seeing that it required 2 counts to get to 6. This strategy, along with counting backward from the first number, is rare for kindergartners, however.

No Observable Strategy

The child is not observed using a strategy. He or she simply stares or looks down and after some time, gives an answer. This is different from "quick retrieval" in that there is wait time before the answer is given.

Sample Cases of Strategy Use

If a student uses more than one strategy to arrive at the answer, the examiner marks all strategies used appropriately. The following are some examples of children's responses for the problem 3 + 2 and how they should be coded:

Example 1

A child draws 3 circles, then 2 circles. The child then counts all the circles to obtain the sum, 5. Circle *drawing* and *counting—all*.

Example 2

A child holds up 3 fingers on one hand and 2 fingers on the other hand, then proceeds to count all the fingers beginning with "1." Circle *fingers* and *counting—all*.

Example 3

A child points to 3 on the number list, then counts up 2 spots, "4, 5." Circle *number list* and *counting—on*.

Example 4

A child draws 3 circles to represent the first addend. The child then points to the 3 circles, says "3" and then counts "4, 5" while putting up 2 fingers (one at a time). Circle *drawing, fingers,* and *counting—on*.

Example 5

A child, after hearing a story problem, writes the combination 3 + 2 on the paper. He or she then holds up 3 fingers and 2 fingers and counts "1, 2, 3, 4, 5." Circle *written combination, fingers,* and *counting—all*.

Scoring the Number Sense Screener

Scoring the NSS is straightforward. One point is given for each correct response, with a total of 29 possible points. The NSS quick script gives detailed instructions for how to score each item, but examiners should be fully acquainted with this section to be sure of accurate scoring methods.

At the conclusion of the test, the examiner performs the following calculations to arrive at a score:

1. Calculate the subarea raw scores. Each subarea is scored separately and recorded on the NSS record sheet under Subarea Totals. These scores can be used for qualitative analysis of the child's strengths and weaknesses and are helpful for planning instruction.

2. Calculate the total NSS raw scores. Find the total NSS raw score by adding the subarea totals, and enter the total score in the appropriate space on the NSS record sheet.

3. Convert raw scores to standard scores and percentile ranks. After the subarea and total raw scores have been calculated and entered in the appropriate blanks on the NSS record sheet, they must be converted to standard scores and percentile ranks. These scores can be obtained by referring to the standard scores and percentile ranks tables in Appendix D of this user's guide. Choose the table appropriate for the time you administered the NSS. Identify the raw score on the left of the appropriate table, and then read across for the standard score and the percentile rank. Write the standard score and the percentile rank in the appropriate space on the NSS record sheet.

CHAPTER 3

Interpreting and Using the Number Sense Screener™ ⠋nss™

The NSS™ total raw scores have been converted to standard scores and percentiles to allow for comparison with other standardized measures. The standard scores have a mean of 100 and a standard deviation of 15. Percentiles are obtained from standard scores and allow for comparison to the norm group. Assuming a normal distribution, a percentile score of 75 means that the student has performed better than 75% of the norm group at the same time of testing. (Standard scores and corresponding percentile ranks should be interpreted cautiously, because children from the normative sample came from a single geographic region.)

Using Tables 1–3 in Appendix D, the child's percentile rank can be determined from the NSS total raw score (see Chapter 2). If the NSS is being used as a screener in the fall of kindergarten, it is important to look for children who fall below the 25th percentile, as they are considered at risk for mathematics difficulties and should be considered for early intervention (see Table 3.1 for raw scores and percentiles). The 25th percentile does not always fall on a particular score, but rather between two scores. To err on the side of caution, the higher of these two scores is listed. It is also helpful to look at the 50th percentile, as it gives an indication of average performance.

When testing in the fall or spring of kindergarten, the NSS can be used to predict possible difficulties in first grade and suggest areas in need of intervention during kindergarten. As can be seen by the jump in the raw score for the 25th percentile rank from the spring of kindergarten to the fall of first grade, there should be considerable growth in number sense from the spring of kindergarten through the first few months of first grade. This sort of growth can occur only if the child has a strong foundation. It is helpful to look for specific weaknesses in number sense that may make this growth difficult.

The NSS can also be used as a screener in the fall of first grade to detect any child at risk for math difficulties, especially if the child has not been screened in kindergarten. Such screening allows for early intervention in first grade, before the child begins to fall farther behind.

Qualitative Inspection of Subarea Scores and Individual Items

As with any raw test score, the Total NSS Score tells about a child's overall number sense, but it does not tell about his or her particular strengths and weaknesses. The NSS is uniquely designed to provide information beyond a standardized score. To devise a plan for intervention, it is necessary to look more closely at a child's performance in each subarea. The examiner should look for inconsistencies between and among subareas. For example, a child may count

Table 3.1. Raw scores for determining the 25th and 50th percentile rank in Number Sense Screener

Time of testing	Raw score (percentile)	Raw score (percentile)
Fall of kindergarten	12 (32nd)	15 (53th)
Spring of kindergarten	13 (30th)	17 (55th)
Fall of first grade	20 (25th)	25 (50th)

accurately to 30 but may not recognize any number over 4. If a child can solve story problems but not number combinations, why might that be?

This section presents an overview of each subarea, accompanied by possible interpretations of student performance in each. It is important that the interpreter of the assessment be thoroughly acquainted with the administration and scoring of the NSS (see Chapter 2).

Counting Skills

The Counting Skills items are typically easy for a kindergartner in late fall. The test was designed to begin with items that are easy for most kindergarten children. However, this also means that if a child has difficulty with these tasks, intervention is needed at a very basic level. The first task (item 1) in Counting Skills is used to assess whether a child can enumerate a small collection using one-to-one correspondence. Failure to complete this task correctly indicates a need for help with basic counting skills. Intervention should include tasks in which the child counts both physical objects (e.g., chips) and pictures of objects (e.g., dots). It is usually best for the child first to count horizontal arrays from left to right with homogeneous collections. Then the child can learn to count different sets presented in any order.

Item 2 assesses whether the child has mastered the *cardinality principle* (i.e., understanding that the last number he or she says when counting connotes the number of objects in a set). Although items 1 and 2 are closely related, it is possible for a child to know how to count collections without understanding the cardinality principle; although the child can enumerate the objects, he or she does not understand that the last number said is the answer to "how many?" Intervention should include enumeration of objects as with item 1 but must include the question "how many?" For example, the teacher may put out eight chips and ask the child to count them. After the child counts, the teacher would ask, *"How many are there?"* Children who have not mastered the cardinality principle often respond by counting the objects again. The teacher should demonstrate by putting out eight chips and saying, *"Watch me: one, two, three, four, five, six, seven, eight,"* while touching each one. Then the teacher should draw an imaginary circle around the entire set with his or her finger and say, *"Eight. How many chips?"* He or she should circle again and say, *"Eight."*

The next level of intervention would include covering the objects after the child has enumerated them and asking again, *"How many?"* As before, the teacher can demonstrate appropriate counting and responses. The child also could simply be asked to give the teacher a specified number of objects (e.g., *"Give me six chips"*).

Item 3 determines whether a child can count to 10 orally. Some children have mastered rote oral counting but not one-to-one correspondence or cardinality. That is why the first two counting items are so important. A teacher may misinterpret accurate oral counting as indicating a mastery of other counting principles. However, it is possible that a child is simply reciting words like a child recites the alphabet, with no firm understanding of the quantities the number words represent.

If the child is an English language learner, it should be determined whether the child can count and enumerate in his or her first language. This would indicate whether the problem is mainly a language problem rather than a counting problem.

Number Recognition

Most children can recognize the numbers 1–10 by late fall of kindergarten, so numbers below 10 were not found to be helpful in predicting future success in mathematics (Jordan, Glutting, Ramineni, & Watkins, 2010). However, having trouble recognizing the numbers 1–10 can indicate a serious delay. Although in the Number Recognition subarea the numbers 2, 4, and 9 are considered a warm-up and are not included in the Total NSS Score, teachers should take note if a child misses any of these three numbers.

The numbers 11–19 can be particularly difficult for children to name, because the English system of naming these numbers is inconsistent in several ways. First, it is inconsistent with the system of naming the other two-digit numbers. It would be more reasonable if the numbers 11–19 were named ten-one, ten-two, ten-three, and so on. Second, it is inconsistent within itself in that most are read with the ones digit followed by "teen"—except for 11, 12, 13 (thirteen, not three-teen) and 15 (fifteen, not five-teen). Also, 11 and 12 have unique names that do not follow even this pattern. It would make more sense if 11 were "one-teen" and 12 were "two-teen." Because of this difficult naming system, children may require extra practice with naming these numbers.

Numbers greater than 20 are easier to recognize if the child knows the base-ten system of naming these numbers. Many kindergarten curricula do not explicitly teach two-digit numbers as a group of tens and ones. Research (Dyson et al., in press; Fuson, Grandau, & Sugiyama, 2001) suggests that kindergartners can benefit from such instruction, which allows them to generalize to numbers they have not practiced. If by the spring of kindergarten, children are not able to name two-digit numbers, an intervention explicitly addressing the naming system as groups of tens and ones may be helpful. It also is helpful for children to learn the decade words (e.g., twenty, thirty).

Number Comparisons

The Number Comparisons subarea addresses the child's ability to compare number quantities when presented in symbolic form, both verbally and numerically. Many number skills are based on the child having an understanding of the linearity of the counting numbers, making it an important number sense idea (Baroody et al., 2009; Baroody, Ginsburg, & Waxman, 1983). Knowing that each number is one more than the number before it and one less than the number after it, no matter how big the number, allows for simple addition and subtraction problems to be solved by moving up and down the number list. It also allows children to make magnitude comparisons for numbers without seeing the physical quantities. Children come to understand that farther up the number list means a larger quantity. A child's ability to form a linear mental number line has also been associated with mathematics achievement generally (Booth & Siegler, 2006; Geary et al., 2007; Siegler & Booth, 2004).

Items 1 and 2 assess "number after" skills. Some children find the number "right after" easy to say, because it is the number said "after" in the count sequence. However, children have more trouble finding the number "two numbers after," because this task associates a quantity "2" to the concept "after."

Items 3–6 require children to make magnitude comparisons between two numbers without seeing a physical representation of the quantity. Again, this draws on a child's understanding that numbers farther up the number list are greater in quantity than those farther down the number list. Task 7 requires a child to compare the distance a number is from 5.

Children who find these tasks difficult would benefit from activities that use a linear representation of number. Games and activities using a number list (see materials list in Chapter 2) can help children to see how numbers increase by one as they go up the number list and decrease by one as they go down the number list. Many commercially available board games, such as Chutes and Ladders, make use of number lists. (See Dyson et al., in press, and Ramani & Siegler, 2008, for specific intervention ideas.)

Nonverbal Calculation

Nonverbal Calculation tasks assess a child's ability to manipulate quantities without requiring verbal number words. Most kindergarten children find items 1 and 2 easy because they can "see" the answer without counting. The multiple choice answers also support the child in seeing the answer. If the child has trouble with these items, it may suggest fundamental difficulties. The child should be asked to visualize the numbers moving in and out of the box. They also might need practice *subitizing* (i.e., automatically seeing quantities without counting) to 5. Item 3 is more difficult because it requires the child to "see" seven objects if he or she cannot find the answer through counting. This quantity is more than a child is able to subitize. If a child responds correctly to the first two nonverbal items but incorrectly to the third one, it may be the case that the child is "seeing" the quantity and not using a counting strategy. This would hinder the child in solving problems with sums larger than 5. The child also might have basic problems with one-to-one correspondence.

An intervention may include opportunities for the child to manipulate first small and then larger quantities through addition and subtraction. It is important for the child to manipulate the quantities themselves and not just observe the teacher. Tasks might include both activities where the total is not hidden as well as activities where one addend or the total is hidden. This helps children see numbers as being made up of sets of smaller numbers and allows flexibility from addend to sum in the case of addition. Practice with the nonverbal calculation materials also helps children develop counting—on strategies (e.g., recognizing the first number of dots and counting-on the number of dots being put into the box to make the total).

Children's understanding can be more fully understood by comparing their responses to the Nonverbal Calculation tasks with their responses to Story Problems and Number Combinations.

Story Problems

It is sometimes the case that children are successful at the first four subareas on the NSS and then are unable to solve the Story Problems, which do not provide physical object representations. Some children are unable to extract the quantities from the story and then perform the appropriate operation, even though they are able to manipulate numbers when not in context of a story. The following are some common error patterns:

1. Stating as the answer the first number mentioned in the problem
2. Stating the second number mentioned in the problem
3. Stating a same random "favorite number" for all the problems
4. Stating very large numbers that could not possibly result from the action described in the problem

Interventions should focus on making models of story problems. The teacher might read a story to the child and ask him or her to make a drawing of the problem or to make a model using chips. Again, the intervention should begin with the teacher modeling a few problems. Teachers often take for granted that children can set up story problems if given manipulatives, but this is not the case. Children often use manipulatives inappropriately. For the problem *"José has 3 cookies. Sarah gives him 2 more cookies. How many cookies does José have now?"* the child might pull out three chips, then push them back and pull out two chips and say "two" for the sum. Explicit instruction in modeling story problems has been shown to be an effective intervention for kindergarten children (Dyson et al., in press). If children have been unsuccessful in the earlier subareas, those skills should be built first before working on story problems.

There is an important connection between a story problem and the number combination that it represents. Children should practice linking story problems to their respective combinations. For example, for the story problem *"José has 3 cookies. Sarah gives him 2 more cookies. How many cookies does José have now?"*, children might be asked to write the combination that represents this problem (3 + 2) or choose the combination that represents this problem from a set of combination cards.

Number Combinations

Unlike the Story Problems subarea, the Number Combinations subarea does not require the child to extract the quantities from a story. Also, the written number combinations are presented in the *Number Sense Screener™ (NSS™) Stimulus Book, K–1, Research Edition,* whereas nothing is presented to the child for Story Problems. Children typically have a harder time with these problems than with story problems, because there are no references to objects. Research (Dyson et al., in press) suggests that often children believe they must just "know" the answer and that there is no way of figuring it out. Children who used strategies for the Story Problems subarea often drop these strategies for Number Combinations and respond with such phrases as "My mom taught me these" or "I don't know that one!" When it comes to a combination that they have not memorized, they are unable to find the answer another way (e.g., counting on fingers, making a drawing). Interventions would be similar to those mentioned in the earlier discussion of Story Problems, but emphasis should be on presenting the written combination and having the children make a model and/or use their fingers to solve the problem. Children should not simply be taught to memorize facts without understanding of the operation.

Strategy Use

Some children have had more experience than others in using adaptive strategies for calculating. Whereas some children have been taught to count on their fingers or to count out objects to solve a problem, others try to solve problems by guessing or relying on their memory. A child who draws a picture to represent a story problem that asks him or her to calculate 4 + 2 and loses count, resulting in an answer "5," tells much more about his or her thinking than a child who simply says "five" for a reason unknown. In the same way, a child who guesses an answer and makes no attempt to use tools provided to him or her reveals that he or she needs more experience in modeling problems and making connections between problems and quantities.

Case Studies

To further aid in interpretation, three case studies that reflect common patterns of performance are presented. The cases are not real children but rather represent some common cases that have been observed. Accompanying these case studies is an interpretation of the performance and suggestions for intervention. As was shown in the previous section, number sense intervention consists of much more than rote learning of numbers and number combinations.

MARIA

In the fall of kindergarten, Maria came happily to testing and seemed happy to spend some one-on-one time with the examiner. In the Counting Skills subarea, she counted the five stars correctly, but when asked, *"How many stars were on the paper you just saw?"* she said "four." When asked to count, she counted perfectly to 10 but could not count any higher. For Number Recognition, she could recognize the warm-up numbers 2 and 4 but could not name 9 or any of the test items (13, 37, 82, 124). In Number Comparisons, she named the number "right after 7" but not the number "two after 7." She responded correctly to one item in the bigger/smaller Number Comparisons items, but as it is an either/or choice, it could have been by chance.

Maria got three of four questions correct in the Nonverbal Calculation subarea, missing 4 + 3. She scored a zero for both the Story Problems and the Number Combinations subareas. She used no overt strategies and had two distinct error patterns. For the Story Problems she said "nine" for every problem, and for the Number Combinations she verbally combined the two numbers (for 2 + 1 she said "two-one"). Her NSS total score was 7, which puts her at the third percentile for fall of kindergarten.

Maria is a classic case of a child who needs support in number sense. Although she has learned the count sequence and can enumerate items, she has not mastered cardinality. Because she got items 1, 2, and 4 correct in the Nonverbal Calculation subarea, Maria seems to be able to subitize small quantities but is unable to solve "4 + 3" using her visual skills only. Intervention with Maria should begin with counting skill activities, then progress to number list activities. Number recognition over 10 can begin after she has mastered the numbers 1–10. More advanced story problem and number combination activities should be held back until she has mastered the numbers/quantities 1–10 and moving forward and back on the number list.

ERIC

Eric arrived at testing eager to participate in the assessment. He informed the examiner that he was "good at math." He easily completed the counting tasks and readily counted to 20 without error. He could name 13, 37, and 82 but missed 124. Eric performed well on Number Comparisons, missing only "two after 7." While working through the Nonverbal Calculation subarea, he remarked, "This is fun!" His only error was for 4 + 3.

When Eric reached the Story Problems subarea, he was able to answer the first problem, *"Jill has 2 pennies. Jim gives her 1 more penny. How many pennies does Jill have now?"* He quickly responded "three." Then everything fell apart for Eric. His response to item 2, *"Sally has 4 crayons. Stan gives her 3 more crayons. How many crayons does Sally have now?"* was "4." His response to item 3, *"José has 3 cookies. Sarah gives him 2 more cookies. How many cookies does José have now?"* was "3." This pattern continued for both subtraction problems. He simply named the first number mentioned in the story. His responses to the Number Combinations subarea was the same. He only responded correctly to "2 + 1." It is important to note that Eric used quick retrieval for the two problems he got correct and did not use any strategies for the remaining problems.

It appears that Eric has a good memory and has a sense of the relative size of numbers. The fact that he was able to find "2 + 1" in three different contexts, both with and without a model, indicates that he knows that "plus one" gives the next number. However, it appears that he does not understand combining two smaller sets to make a larger set when a physical model is not given. He might have been able to "see" 3 + 2 in the Nonverbal Calculation subarea and match it to a picture, but when he was presented with the Story Problems or the Number Combinations, he had no tools with which to solve the problems.

Intervention for Eric would not be necessary if this were his performance at fall of kindergarten. It is likely that kindergarten math activities would be enough for Eric to learn problem-solving skills. He has the foundations for learning to add and subtract. However, if this were the spring of kindergarten, or the fall of first grade, there would be reason for concern. Eric shows little ability to manipulate small sets of numbers—a skill so important for being successful in elementary mathematics; neither does he have any tools available to him to help him solve these problems. In these cases, the intervention activities listed earlier under Story Problems and Number Combinations would be appropriate for Eric.

CELINE

Celine is in a bilingual kindergarten classroom. Spanish is the primary language spoken by her parents. She came to testing with a shy and quiet demeanor. She worked diligently throughout the assessment but had trouble with individual items throughout the various subareas. The following is a summary of her scores on the subareas:

Subarea Totals

A	Counting Skills Total: 3/3	D	Nonverbal Calculation Total: 3/4
B	Number Recognition Total: 1/4	E	Story Problems Total: 0/5
C	Number Comparisons Total: 4/7	F	Number Combinations Total: 6/6

Celine scored perfectly on Counting Skills and could recognize 2, 4, and 9 (the warm-up numbers) but could only recognize the number 13 out of the four test items. For Number Comparisons, she responded incorrectly to three out of the four bigger or smaller items but knew the number "after 7" and "two after 7." Celine missed the first Nonverbal Calculation problem only, and as the other items were harder, it is safe to assume that she made a careless error.

The next two subareas help to tell the story. Celine did not respond correctly to any of the Story Problems, nor did she use any obvious strategies. However, she responded perfectly to the Number Combinations, using her fingers to solve each one. Again, if the time of testing was fall of kindergarten, the teacher may think that there is no cause for alarm and the teacher might be careful to make sure that Celine is making progress when story problems are presented in class. Although having difficulty with story problems is common early on in kindergarten, being successful with number combinations is not.

In this case, it is important to look at the discrepancy between the story problems and number combinations. Because Celine is proficient at combinations, the evaluator should wonder why she has trouble with story problems. Is it that she cannot extract the numbers from the story or that she is an English language learner and is having trouble understanding the problems? Does she have a language disability that makes it difficult for her to process story problems? Further language ability testing would be helpful in determining appropriate intervention.

Conclusion

Successful interpretation of NSS results requires the examiner to become very familiar with the components of number sense, their similarities and differences. The NSS gives considerable information beyond a standard score or percentile rank. Carefully studying Chapters 2 and 3 before administering, scoring, and interpreting the NSS will result in more effective use of this tool.

CHAPTER 4

Reliability and Validity of the Number Sense Screener™ ⁛∷nss™

The establishment of a test's reliability and validity is an ongoing endeavor, and it is one that extends well beyond a test's development and standardization (Gregory, 2007). Consequently, it is impossible to present a complete set of psychometric characteristics at the time of a test's publication. This chapter presents current reliability and validity data for the NSS™. The authors, along with independent researchers, will continue to provide new technical information on the NSS as it becomes available. Independent researchers, and other interested users, are encouraged to contact the authors regarding the development and/or outcomes of such studies.

Reliability

Reliability is defined as the consistency of a measure's scores across items and across time (Anastasi & Urbina, 1997; Salvia, Ysseldyke, & Bolt, 2009). Several statistics are useful to describe a test's reliability. Among them are person- and item-separation indices obtained from modern test-score theory. Likewise, the provision of internal-consistency reliabilities from traditional test-score theory, standard errors of measurement, and test–retest stability coefficients are all recommended by *The Standards for Educational and Psychological Testing* (American Educational Research Association, American Psychological Association, & National Council on Measurement in Education, 1999).

Item Statistics[1]

This section explains the person- and item-separation reliabilities that are important indices in a Rasch analysis (Bond & Fox, 2007; Rasch, 1960) The Rasch analyses were completed using Winsteps (Linacre, 2007) and SAS 9.1 software. The analyses were directed to children who participated in the item tryout phase of the NSS's development (*N* = 425). These children completed 26 of the 29 items in the final version of the NSS; the three items were not included because they were administered only to children in the NSS's oldest norm group.

The Rasch model was applied because of its desirable properties of linear, interval measurement (Embretson & Reise, 2000) It is important to note that the model provides indices of fit to test model assumptions of appropriate ordering of items and persons, along with issues

---

[1]The Rasch item analyses were completed by Jonathan Rubright, doctoral candidate at the University of Delaware. We are deeply indebted to him for his expertise and assistance.

surrounding dimensionality (Wright & Stone, 1979). Results will be presented using infit and outfit mean squares, z standardizations, difficulty (location) parameters, standard errors expressed in logits, and person- and item-separation reliabilities (Wright & Stone, 1979). Items with mean squares (i.e., the average squared residual) above 1.4 were considered to show misfit. Lower values indicated an item was performing in expected ways in relation to the other scale items. Standardized values should generally be below 2.0 (Wright & Linacre, 1994).

Item and person reliabilities refer to the replicability of ordering: how well items would be similarly ordered given to a new sample or how well persons would be ordered given alternative yet similar items (Wright & Masters, 1982). In addition, a principal components analysis (PCA) was performed on the model residuals to search for extraneous factors (Linacre, 1998). Lastly, a differential item functioning (DIF) analysis was performed to search for items behaving differently across gender. The Mantel-Haenszel contingency table approach was used to identify items displaying DIF (Mantel & Haenszel, 1959). Magnitude of DIF was assessed using the Educational Testing Service (ETS) difficulty delta index, $D = -2.35 \ln (\alpha_{MH})$ (Dorans & Holland, 1993). Cut points were as follows: Class A (negligible DIF) = $|D| < 1.00$, Class B (moderate DIF) = $1.00 \leq |D| < 1.50$, and Class C (large DIF) = $|D| \geq 1.50$.

Results revealed that item-reliably index was .99 and the person-reliability index was .84, providing evidence of reliability (person index) and validity (item index) of the scale. The variance explained by the measure of interest was 83.4%, with only 7.3% of the variance explained by the first factor of residuals. Fit statistics from the analysis are displayed in Table 4.1 for every

Table 4.1.   Number Sense Screener item fit statistics

Item	Location	Location SE	Infit MNSQ	Infit ZSTD	Outfit MNSQ	Outfit ZSTD
1	−3.97	.3	0.85	−0.5	0.43	−1.4
2	−2.34	.17	0.96	−0.4	0.9	−0.2
3	−2.43	.17	0.79	−1.9	0.42	−2.4
4	0.32	.12	0.86	−2.8	0.81	−2.1
5	−0.83	.12	0.91	−1.5	1.23	1.5
6	1.28	.13	1.31	4.5	1.37	2.8
7	−1.87	.15	1.02	0.3	1.28	1.1
8	−0.6	.12	1.23	4	1.35	2.4
9	−0.81	.12	1.14	2.4	1.29	1.8
10	−0.57	.12	1.05	1	1.33	2.3
11	−0.53	.12	1.18	3.3	1.34	2.4
12	−2.55	.18	0.94	−0.4	1	0.1
13	1.13	.12	1.1	1.6	1.1	0.9
14	0.25	.12	0.89	−2.2	0.84	−1.8
15	−1.62	.14	0.98	−0.2	0.93	−0.2
16	0.69	.12	1.08	1.5	1.07	0.7
17	1.94	.14	0.94	−0.7	0.9	−0.6
18	1.25	.13	0.92	−1.4	1	0
19	1.85	.14	1.3	3.8	2.17	5.7
20	1.02	.12	0.96	−0.6	1	0
21	0.24	.12	0.85	−3	0.82	−1.9
22	1.16	.12	0.8	−3.5	0.87	−1.2
23	1.53	.13	0.77	−3.7	0.72	−2.3
24	1.04	.12	0.85	−2.8	0.8	−2
25	1.92	.14	1.01	0.2	1.11	0.7
26	2.5	.15	0.81	−2.1	0.56	−2.4

Note: N = 425.

Key: location, location (difficulty) parameter; location SE, location standard error; infit MNSQ, infit mean squares; infit ZSTD, standardized infit; outfit MNSQ, outfit mean squares; outfit ZSTD, standardized outfit.

Table 4.2. Number Sense Screener Mantel-Haenszel bias analysis across gender

Item	MHχ^2	p-value	α_{MH}
1	0.85	.36	0.45
2	0.67	.41	1.35
3	1.75	.19	1.67
4	3.17	.08	1.57
5	0	.99	1.00
6	0.07	.80	1.07
7	0.03	.87	1.05
8	0.04	.84	1.05
9	0	.98	1.01
10	0.04	.84	1.05
11	1.23	.27	1.29
12	2.65	.10	0.53
13	0.24	.63	0.89
14	2.86	.09	1.57
15	0.04	.85	1.06
16	2.92	.09	0.65
17	0.68	.41	0.77
18	0.24	.63	0.87
19	0.50	.48	1.20
20	0.16	.69	1.11
21	2.25	.13	0.67
22	1.27	.26	1.42
23	0.34	.56	0.83
24	6.08	.01	0.50
25	0.51	.48	0.82
26	0.76	.38	1.40

Note: MHχ^2 = Mantel-Haenszel summary chi-square test, α_{MH} = Mantel-Haenszel alpha. $N = 425$.

item. Outcomes from the analysis serve to support the NSS's construct validity by showing that the instrument essentially measures one construct.

Table 4.2 presents results from an item-bias analysis. The analyses were completed using Mantel-Haenszel methodology; it compared genders and used males as the reference group. Results revealed that only one item exhibited significant DIF. Specifically, girls answered item 24 correctly about twice as often as boys ($1/.50 = 2$). In log odds terms, girls found this question to be $\beta = \ln(.50) = -.69$ times easier than boys with equal ability. Using the ETS index, the magnitude of the difference is $D = -2.35 \ln(.50) = 1.62$, a large difference. Nevertheless, only one instance of bias was found across 26 items. Therefore, it is fair to infer that the NSS is essentially free of gender bias.

Internal Consistency

Cronbach's (1951) coefficient alpha was used to calculate internal-consistency reliability. Coefficient alpha provides a lower bound value of internal consistency and is considered to be a conservative estimate of a test's reliability (Allen & Yen, 1979; Carmines & Zeller, 1979; Reynolds, Livingston, & Willson, 2006). Alpha coefficients are presented in Table 4.3 for each of the NSS's three norm groups. Coefficients are also presented separately for males and for females. Lastly, the bottom row of Table 4.3 presents averaged values.

Table 4.3. Internal-consistency reliability for the Number Sense Screener

	Demographic cohort		
Norm group	Total sample[a]	Males	Females
Fall of kindergarten	.82	.83	.82
Spring of kindergarten	.86	.89	.85
Fall of first grade	.87	.87	.87
Average[b]	.85	.87	.85

[a]$N = 425$.

[b]Average coefficients were calculated with Fisher's z' transformation.

Table 4.3 demonstrates that, on average, reliabilities increased with children's age. For instance, internal consistency for the total sample was .82 for youngest children who were assessed during the fall of kindergarten. It continued to increase for children evaluated during the spring of kindergarten (.86) and reached its zenith at .88 for children evaluated during the fall of first grade. The average value for the entire sample was .85. As expected, this internal-consistency reliability coefficient aligned well with the person reliability index (.84) from the Rasch analyses.

Averaged values at the bottom of Table 4.3 are appropriately high for the entire sample, and for males and females (all three equal .85 to .87) and exceed the .80 criterion suggested in certain textbooks for achievement measures (e.g., Reynolds et al., 2006). Consequently, results indicate examiners can use the NSS with confidence, because its scores demonstrate high levels of internal-consistency reliability.

Confidence Intervals

The presence of random error ensures that no test can be perfectly reliable. That is, scores of a child who is retested on different occasions with the same instrument or retested with a different set of equivalent items from the same instrument vary somewhat. The most common method of indicating unreliability is to supply confidence intervals for scores.

The standard error of measurement (SEM) provides the foundation upon which confidence intervals are built. The SEM is reflected whenever the 68% confidence level is reported for a score, because the SEM and the 68% confidence limit are the same (see Glutting, McDermott, & Stanley, 1987, for more information). A 68% confidence level means that over a large number of testings, examiners can be 68% confident that a child's true score resides within a specified score range (Dudek, 1979).

Table 4.4 presents confidence interval magnitudes for standard scores ($M = 100$, $SD = 15$) for the NSS. As the plus (+) symbol indicates, these values should be added and subtracted to children's standard scores in order to establish the upper and lower bounds within which their true scores are likely to fall. In addition to presenting the intervals for the 68% confidence level, Table 4.4 provides the magnitudes of the intervals for other common confidence levels, including 90%, 95%, and 99% levels. All values are reported by NSS norm group.

Table 4.4. Confidence interval magnitudes for Number Sense Screener standard scores

	Confidence level			
Norm group	68%	90%	95%	99%
Fall of kindergarten	±6.4	±10.4	±12.5	±16.4
Spring of kindergarten	±5.6	±9.2	±11.0	±14.5
Fall of first grade	±5.4	±8.9	±10.6	±13.4

Note: Confidence intervals are expressed for standard scores ($M = 100$, $SD = 15$) and based on total sample reliabilities, by age, presented in Table 4.3.

Table 4.5.   Number Sense Screener test–retest reliability coefficients

| Time of administration | Time of administration | | | | | |
	September K	November K	February K	April K	November Gr. 1	February Gr. 1
September K	—	.81	.80	.78	.69	.61
November K		—	.82	.81	.70	.61
February K			—	.86	.77	.70
April K				—	.81	.75
November Gr. 1					—	.80
February Gr. 1						—

From Jordan, N.C., Glutting, J., Ramineni, C., & Watkins, M.W. (2010). Validating a number sense screening tool for use in kindergarten and first grade: Prediction of mathematics proficiency in third grade. *School Psychology Review, 39*(2), 187; Copyright 2010 by the National Association of School Psychologists. Bethesda, MD. Reprinted with permission of the publisher www.nasponline.org.

Note: K, kindergarten; Gr. 1, Grade 1. *N* = 378.

Test–Retest Stability

Score stability was examined by using data from 378 children who took part in a 4-year longitudinal investigation of children's mathematics development (Jordan et al., 2006). Participants attended the same public school district in northern Delaware. All kindergartners from six schools were invited to participate in the study. There were 378 children who started the study at the beginning of kindergarten and 204 who remained at the end of third grade. Participant attrition was due to children moving out of the school district (typically right after kindergarten), rather than withdrawal from the study or absence on the day of testing. A logistical regression analysis (Jordan, Kaplan, Ramineni, & Locuniak, 2009) revealed that although gender and age do not predict the odds of being absent from the study in third grade, low-income and minority children, respectively, were about 1.2 times more likely to be absent from the study than were middle-income and nonminority children. In third grade, 52% of the children were boys, 45% had minority ethnic backgrounds (63% African American, 26% Hispanic, and 11% Asian), and 23% came from low-income families. Income status was determined by participation in the free or reduced-price lunch program in school, and most low-income children resided in urban neighborhoods.

Table 4.5 presents test–retest reliability coefficients across the six time periods. As expected, stability coefficients are higher for shorter intervals. Reliabilities ranged from .61 to .86. Twelve of the fifteen reliability coefficients were at or above the .70 criterion recommended in certain assessment textbooks (Gregory, 2007; Reynolds et al., 2006). Three coefficients dipped below the .70 criterion. However, this occurred only when the testing period exceeded 1 year. Findings therefore point to the need for annual retesting with the NSS.

Validity

Validity is multifaceted (American Educational Research Association et al., 1999); yet, at its core, the construct can be viewed simply as the extent to which a test measures what it is designed to measure (Gregory, 2007; Salvia, Ysseldyke, & Bolt, 2009). A variety of statistical data was gathered to document the relevance of information provided by the NSS. The NSS's validation strategy is consistent with the substantive-construct model of test development, wherein a test's validity is examined both internally to itself and externally to criterion variables (cf. Cronbach & Meehl, 1955). Consequently, the presentation below is divided into six sections according to the types of evidence testable at the time of the NSS user's guide's publication: 1) developmental changes, 2) content-related validity, 3) discriminant (contrasted-groups) validity, 4) predictive validity, 5) construct validity, and 6) treatment validity.

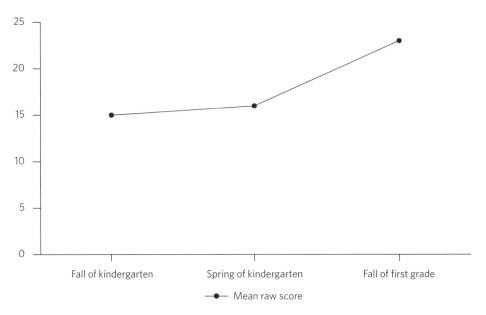

Figure 4.1.   Plot of mean Number Sense Screener raw scores across age.

Developmental Changes

Age differentiation is a major criterion employed in the validation of a number of intelligence tests and achievement tests (Anastasi & Urbina, 1997). Because mathematics knowledge is expected to increase with age during childhood, it is argued that valid tests show raw scores that increase with age. Figure 4.1 plots mean raw scores from the NSS's three norm groups. The plot reveals that the NSS possesses considerable developmental validity, because raw scores exhibit clear-cut and consistent age changes.

Content-Related Validity

The content of the NSS is well established by the research (Jordan et al., 2010). The assessment is closely aligned to the Kindergarten Common Core State Standards in Number: Counting and Cardinality, and in Numbers and Operations in Base 10 (National Governors Association Center for Best Practices & Council of Chief State School Officers, 2010). According to the Common Core, kindergartners are expected to know number names and the count sequence, to count to tell the number of objects in a collection (NSS Counting Skills subarea), and to compare numbers presented as written numerals between 1 and 10 (NSS Number Comparisons subarea). In terms of operations, kindergartners are expected to understand addition as adding to a quantity and understand subtraction as taking from a quantity (NSS Nonverbal Calculation, Story Problems, and Number Combinations). Kindergartners also are expected to work with numbers between 11 and 19 to gain foundations for place value (NSS Number Recognition).

The NSS also lines up with the Kindergarten Focal Points of the National Council of Teachers of Mathematics (2006) in the area of numbers and operations. This includes representing, comparing, and ordering whole numbers (NSS Counting Skills, Number Recognition, and Number Comparisons) and joining and separating sets (NSS Nonverbal Calculation, Story Problems, and Number Combinations).

Discriminant (Contrasted-Groups) Validity

Campbell and Fiske (1959) introduced the concept of discriminant validity. They stressed that new tests need to be evaluated using both discriminant and convergent validation techniques. One way to establish a discriminant validity is to determine how well scores from a test distinguish (i.e., discriminate) children who meet achievement standards on an external measure from children who fail to meet standards.

Participants

Participants were part of a 4-year longitudinal investigation of children's mathematics development (Jordan et al., 2006). This sample was described previously in the section on test–retest reliability ($N = 378$). Readers are referred to that section for further details about the sample's composition.

Criterion

Mathematics achievement was assessed with the third grade version of the Delaware Student Testing Program (DSTP) in Mathematics (Delaware Department of Education, 2008). The DSTP measures concepts and procedures in accordance with Delaware mathematics standards (i.e., numeric reasoning, algebraic reasoning, geometric reasoning, and quantitative reasoning). It has strong internal reliability (.93) and has established cut scores for meeting state standards and for performing below state standards. The test's cut points and content were fully validated by a panel of experts (Delaware Department of Education, 2008). The DSTP in third grade is highly correlated ($r = .77$, $p < .01$) with scores on the Woodcock-Johnson III—Mathematics (McGrew, Schrank, & Woodcock, 2007), indicating strong criterion validity (Jordan et al., 2009). For the present study, the DSTP mathematics outcome measure was used in categorical form (1 = met standards, 0 = did not meet standards). The original five performance levels were collapsed to two levels to simplify the measurement scale. The performance levels of 3 (*meets the standard*), 4 (*exceeds the standard*), and 5 (*distinguished performance*) were transformed to a 1 on the categorical scale to represent meeting the DSTP standards, whereas the remaining two lower performance, levels 1 (*well below the standard*) and 2 (*below the standard*), were transformed to a 0 on the categorical scale to denote failure to meet the standards on DSTP in mathematics.

Procedure

The NSS was given to children individually in school by one of several trained graduate or undergraduate research assistants. It was administered in September and April of kindergarten and in November of first grade. The state mathematics proficiency achievement test was group-administered by school personnel in April of third grade.

Results

Data were analyzed using a repeated measures analysis of variance (ANOVA). Time was the within-subjects (repeated) measure, and it was evaluated on three occasions: fall of kindergarten, spring of kindergarten, and fall of first grade. Table 4.6 presents means and standard deviations for the groups on the dependent variable, and it does so separately by time period. Mauchly's test indicated that the assumption of sphericity was not violated ($\chi^2 = 2.459$, $df = 2$, $p = .001$). Therefore, there was no need to correct for a sphericity violation.

Figure 4.2 provides a visual representation of the results. It shows children meeting proficiency on the DSTP in third grade had higher NSS scores in the fall of kindergarten than children who did not meet proficiency. This situation occurred again in the spring of kindergarten, and remained true in the fall of first grade. Given this arrangement of scores, it came as no

Table 4.6. Means and standard deviations for Number Sense Screener scores by group and time

Time	DSTP criterion group	M^a	SD
Fall of kindergarten	Failed to meet proficiency	89.1	11.5
	Met proficiency	107.0	14.6
Spring of kindergarten	Failed to meet proficiency	86.4	11.1
	Met proficiency	106.6	15.1
Fall of first grade	Failed to meet proficiency	90.3	10.2
	Met proficiency	105.2	12.4

From Jordan, N.C., Glutting, J., Ramineni, C., & Watkins, M.W. (2010). Validating a number sense screening tool for use in kindergarten and first grade: Prediction of mathematics proficiency in third grade. *School Psychology Review, 39*(2), 188; Copyright 2010 by the National Association of School Psychologists. Bethesda, MD. Reprinted with permission of the publisher www.nasponline.org.

Note: DSTP, Delaware State Testing Program; *M*, mean; *SD*, standard deviation; mg, milligram.

^aAll numbers rounded at first decimal point for convenient presentation.

surprise that neither was the main effect for time significant ($F = .783$, *df* [2, 266], $p = .001$) nor was the group x time interaction significant ($F = 2.071$, *df* [2, 266], $p = .458$). Specifically, it was anticipated that a statistically significant main effect would occur only for two groups. In fact, the result revealed that the main effect for group was significant ($F = 39.812$, *df* [1, 133], $p = .001$). This *F*-test evaluates whether the dependent variable changes across groups—independent of time (Keppel & Wickens, 2004). So, in the current case, the main effect for time shows that across each of the three time periods (fall of kindergarten, spring of kindergarten, fall of first grade), children meeting proficiency on the DSTP had higher NSS scores.

Partial eta square (η^2) is a common effect-size measure for repeated measures ANOVA. Murphy and Myors (2004) defined the ranges of partial eta square: .01 = a small effect size, .06 = a medium effect size, and .14 = a large effect size. In the current study, the main effect for group represented a very large effect size (i.e., partial eta squared = .23). Consequently, it is reasonable to infer that the NSS possesses substantial discriminant (contrast-group) validity.

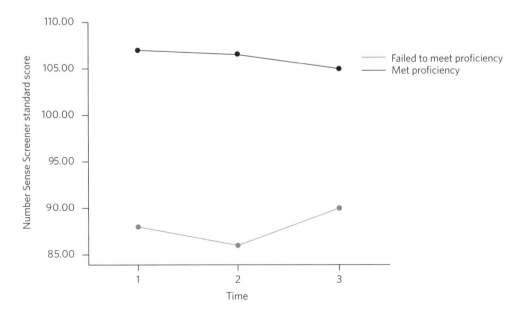

Figure 4.2. Plot of marginal means by group across time.

Table 4.7. Demographic information for participants at the end of first grade ($n = 279$) and the end of third grade ($n = 175$)

Variable	End of first grade	End of third grade
Gender		
Male	55%	54%
Female	45%	46%
Race		
Minority[a]	52%	42%
Nonminority	48%	58%
Income		
Low income	28%	22%
Middle income	72%	78%
Mean kindergarten start age (SD)	5 years 6 months (4 months)	5 years 6 months (4 months)

Reprinted from *Learning and Individual Differences, 20,* Jordan, N.C., Glutting, J., & Ramineni, C., The importance of number sense to mathematics achievement in first and third grades, 82–88, (2010) with permission from Elsevier.

[a]Minority refers to African American (29%, $n = 81$), Asian (6%, $n = 17$), and Hispanic (17%, $n = 47$) at the end of first grade; and African American (25%, $n = 44$), Asian (6%, $n = 11$), and Hispanic (11%, $n = 19$) at the end of third grade.

Predictive Validity

Children were given the NSS at the beginning of first grade, and mathematics outcomes were obtained at the end of both first and third grades (Jordan, Glutting, & Ramineni, 2009). Outcomes included overall mathematics achievement as well as subareas of written computation and applied problem solving. It was hypothesized that number sense proficiency may be more relevant to applied problem solving than written computation, which may be more dependent on learned algorithms. To examine the unique contribution of number sense (as measured by the NSS) to these later mathematics outcomes, we also added the common predictors of age, verbal and spatial abilities, and working memory skills in our analyses.

Participants

Participants were drawn from a multiyear longitudinal investigation of children's mathematics development (Jordan, Kaplan, Ramineni, & Locuniak, 2009). They all attended the same public school district in northern Delaware. Background characteristics of children in first grade ($n = 279$) and in third grade ($n = 175$) are presented in Table 4.7. The first graders included children who completed all measures in first grade, and the third graders were children who completed all measures in first and third grade. In the first grade sample, 55% of the children were boys, 52% had minority ethnic backgrounds, and 28% came from low-income families. In the third grade sample, 54% of the children were boys, 42% had minority ethnic backgrounds, and 22% came from low-income families. Income status was determined by participation in the free or reduced-price lunch program in school. Participant attrition was due primarily to children moving out of the school district rather than withdrawal from the study or absence on the day of testing.

Procedure

The measures were given to children individually in school by one of several trained research assistants. The NSS items were given in October of first grade, the cognitive measures (Vocabulary, Matrix Reasoning, and Digit Span tests) in January of first grade, and the math achievement measures in April of first grade and again in April of third grade.

Cognitive Tasks

The Wechsler Abbreviated Scale of Intelligence (Wechsler, 1999) was used to assess oral vocabulary and spatial reasoning. A digit span test (Wechsler, 2009) was used to measure short-term

Table 4.8. Correlations between first grade Number Sense Screener and control variables

Variable	Number Sense Screener correlation
Math composite (end of first grade)	.72
Math applications (end of first grade)	.73
Math calculation (end of first grade)	.58
Math composite (end of third grade)	.70
Math applications (end of third grade)	.74
Math calculation (end of third grade)	.66
Kindergarten start age	.19
Vocabulary	.56
Matrix reasoning	.53
Digit span forward	.34
Digit span backward	.50

Reprinted from *Learning and Individual Differences, 20,* Jordan, N.C., Glutting, J., & Ramineni, C., The importance of number sense to mathematics achievement in first and third grades, 82–88, (2010) with permission from Elsevier.

Note: All correlations are significant, $p < .01$.

and working memory. Digit span forward is a measure of short-term recall and digit span backward a measure of working memory or active recall.

Mathematics Achievement

Math achievement was assessed with the Woodcock-Johnson III (Woodcock, McGrew, Schrank, & Mather, 2007). The composite achievement score (math overall) was the combined raw scores for subtests assessing written calculation (written calculations using a paper and pencil format; math calculation) and applied problem solving (orally presented problems in various contexts; math applications).

Results

Raw scores from the NSS were used for all analyses. Bivariate correlations are presented in Table 4.8 between the NSS raw scores and raw scores on the cognitive measures at first and third grades, as well as between the NSS and age at the beginning of kindergarten. All of the correlations were positive and statistically significant (i.e., all p values $< .05$), with the two lowest correlations being kindergarten start age (.19) and digit span forward (.34) and the highest correlations being math applications in first and third grades (.73 and .74, respectively).

A primary purpose of the study was to determine the unique contribution of the NSS in predicting criterion mathematics performance. Specifically, the study examined the extent to which the NSS predicted mathematics performance above and beyond the contribution of the control (nuisance) variables of age and general cognition related to language (vocabulary), spatial ability (matrix reasoning), and memory (digit span forward and digit span backward). To accomplish these goals, students' scores on the NSS were regressed on a series of established mathematics achievement outcomes (math overall, math calculation, math applications) using the two-stage model. At Step 1 (model 1), the control (nuisance) variables entered simultaneously into an analysis. Step 2 (model 2) comprised entry of the NSS. The analyses were used to predict mathematics achievement in first grade and then in third grade. The independent contributions of predictors were evaluated through the interpretation of squared partial coefficients. Effect sizes were estimated for the predictors using Cohen's (1988) f^2, where values of .02 equal a small effect, values of .15 equal a medium effect, and values of .35 equal a large effect.

Table 4.9 presents the results for predicting criterion performance on the mathematics composite score (math overall). Model 1 (age and general cognitive measures) accounted for

Table 4.9. Results of block entry regression for the end of first-grade math overall and the end of third-grade math overall: regression coefficients and variance explained by each block of variables

Model	First-grade math overall					Third-grade math overall				
	B	β	t value	p value	Effect size[a]	B	β	t value	p value	Effect size[a]
One										
Age	.11	.07	1.50	.14		−.07	−.03	−0.55	.59	
Vocabulary	.27	.30	5.58	0		.30	.24	3.62	0	
Matrix reasoning	.34	.31	6.16	0		.56	.39	6.26	0	
Digit span forward	.06	.02	0.33	.74		.68	.15	2.24	.03	
Digit span backward	.78	.23	4.38	0		.99	.17	2.77	.01	
Two										
Age	.05	.03	0.79	.43	—	−.09	−.04	−0.82	.42	—
Vocabulary	.12	.14	2.71	.01	.03	.14	.11	1.78	.08	—
Matrix reasoning	.18	.17	3.62	0	.05	.36	.25	4.27	0	.08
Digit span forward	.11	.03	0.71	.48	—	.39	.09	1.43	.16	—
Digit span backward	.38	.11	2.31	.02	.02	.41	.07	1.26	.21	—
Number Sense Screener	.53	.48	8.97	0	.29	.78	.46	6.83	0	.21

Model	R square	R square change	F change	df 1	df 2	R square	R square change	F change	df 1	df 2
One	.47		47.99[b]	5	273	.45		28.03[b]	5	169
Two	.59	.12	80.43[b]	1	272	.57	.12	46.70[b]	1	168

Reprinted from *Learning and Individual Differences, 20,* Jordan, N.C., Glutting, J., & Ramineni, C., The importance of number sense to mathematics achievement in first and third grades, 82–88, (2010) with permission from Elsevier.

[a]Cohen's (1988) f^2 statistic where .02 is a small effect size, .15 is a medium effect size, and .35 is a large effect size.

[b]$p < .01$.

47% of the variance in math in first grade ($p < .01$), with vocabulary, matrix reasoning, and digit span backward reaching significance, and 45% of the variance in third grade ($p < .01$), with vocabulary, matrix reasoning, digit span forward, and digit span backward reaching significance. Results showed that the NSS made statistically significant, unique contributions to the prediction at first grade ($p < .01$) and third grade ($p < .01$) outcomes in math overall. In each instance, the NSS accounted for about 12% more criterion variance than the control variables. More important, Cohen's (1988) f^2 represented medium to large effect sizes for both first- and third-grade criterion performance (respectively, .29 and .21).

Table 4.10 presents the results for predicting mathematics calculation. Model 1 (age and general cognitive measures) accounted for 35% of the variance in math calculation in first grade ($p < .01$) with vocabulary, matrix reasoning, and digit span backward reaching significance, and 33% of the variance in third ($p < .01$), with vocabulary and matrix reasoning reaching significance. Model 2 accounted for 41% of the variance in first grade, indicating that the NSS measure accounted for 6% more variance than the control variables. Cohen's (1988) f^2 value for the NSS was .10, which represented a small to medium effect size. Results for third grade were more impressive. The NSS accounted for 14% more variance of math calculation than the control variables and Cohen's (1988) f^2 (.26) represented a medium to large effect size.

Table 4.11 presents the results for mathematics applications where the results were most impressive. Model 1 accounted for 44% of the variance in math applications in first grade ($p < .01$), with vocabulary, matrix reasoning, and digit span backward reaching significance, and 45% of the variance in third grade ($p < .01$), with vocabulary and matrix reasoning reaching significance. Not only did the NSS make significant, unique contributions that accounted for 14% to 17% of the criterion's variance, Cohen's (1988) f^2 represented a large effect size in predicting first-grade NSS performance (.44) and third-grade NSS performance (.45). In sum, results show that the NSS possesses substantial levels of predictive validity.

Table 4.10. Results of block entry regression for the end of first-grade math calculation and the end of third-grade math calculation: regression coefficients and variance explained by each block of variables

Model	First-grade math calculation					Third-grade math calculation				
	B	β	t value	p value	Effect size[a]	B	β	t value	p value	Effect size[a]
One										
Age	.07	.02	0.47	.64		.16	.04	0.69	.49	
Vocabulary	.10	.24	3.96	0		.15	.28	3.72	0	
Matrix reasoning	.14	.29	5.19	0		.21	.34	4.87	0	
Digit span forward	−.07	−.05	−0.77	.44		−.08	−.04	−0.57	.57	
Digit span backward	.40	.26	4.38	0		.25	.13	1.73	.09	
Two										
Age	−.01	0	−0.03	.97	—	.03	.01	0.16	.88	—
Vocabulary	.05	.12	2.02	.05	.02	.06	.12	1.69	.09	—
Matrix reasoning	.10	.19	3.40	0	.05	.11	.18	2.75	.01	.04
Digit span forward	−.05	−.04	−0.61	.54	—	−.04	−.02	−0.30	.77	—
Digit span backward	.27	.18	2.97	0	.03	.07	.03	0.51	.61	—
Number Sense Screener	.17	.33	5.19	0	.10	.33	.48	6.86	0	.26

Model	R square	R square change	F change	df 1	df 2	R square	R square change	F change	df 1	df 2
One	.35		29.46[b]	5	273	.33		18.77[b]	5	187
Two	.41	.06	26.89[b]	1	272	.47	.14	47.12[b]	1	186

Reprinted from *Learning and Individual Differences, 20,* Jordan, N.C., Glutting, J., & Ramineni, C., The importance of number sense to mathematics achievement in first and third grades, 82–88, (2010) with permission from Elsevier.

[a]Cohen's (1988) f^2 statistic where .02 is a small effect size, .15 is a medium effect size, and .35 is a large effect size.

[b]$p < .01$.

Table 4.11. Results of block entry regression for the end of first-grade math applications and the end of third-grade math applications: regression coefficients and variance explained by each block of variables

Model	First-grade math applications					Third-grade math applications				
	B	β	t value	p value	Effect size[a]	B	β	t value	p value	Effect size[a]
One										
Age	.37	.09	1.95	.05		.01	0	0.02	.98	
Vocabulary	.17	.31	5.56	0		.22	.28	4.08	0	
Matrix reasoning	.19	.29	5.49	0		.38	.41	6.51	0	
Digit span forward	.12	.06	1.11	.27		.07	.03	0.38	.71	
Two										
Age	.21	.05	1.27	.21	—	−.21	−.04	−0.78	.44	—
Vocabulary	.07	.13	2.54	.01	.03	.08	.10	1.67	.10	—
Matrix reasoning	.09	.13	2.79	.01	.03	.22	.24	4.20	0	.11
Digit span forward	.16	.08	1.65	.10	—	.14	.05	0.89	.37	—
Digit span backward	.11	.05	1.07	.29	—	.04	.01	0.21	.83	—
Number Sense Screener	.36	.52	9.69	0	.44	.55	.54	9.11	0	.45

Model	R square	R square change	F change	df 1	df 2	R square	R square change	F change	df 1	df 2
One	.44		43.01[b]	5	273	.45		30.34[b]	5	187
Two	.58	.14	93.89[b]	1	272	.62	.17	82.97[b]	1	186

Reprinted from *Learning and Individual Differences, 20,* Jordan, N.C., Glutting, J., & Ramineni, C., The importance of number sense to mathematics achievement in first and third grades, 82–88, (2010) with permission from Elsevier.

[a]Cohen's (1988) f^2 statistic where .02 is a small effect size, .15 is a medium effect size, and .35 is a large effect size.

[b]$p < .01$.

Construct Validity

Strong construct validity is suggested whenever there is an appropriate pattern of convergent and divergent associations (American Educational Research Association et al., 1999; Campbell, 1960; Messick, 1989). Convergent validity is demonstrated when a scale correlates highly with other scales with which it shares an overlap of constructs. Discriminant validity is demonstrated when a test does not correlate highly with variables from which it should differ. In the current case, we would expect a pattern of somewhat higher correlations between the NSS and scores from measures of mathematics than between the NSS and scores from measures of reading achievement.

The reason for expecting only "somewhat" higher correlations is because the constructs of reading and mathematics are themselves highly correlated. For instance, the Wechsler Individual Achievement Test–Third Edition (Wechsler, 2009) reports correlations between reading and mathematics composite by grade level from kindergarten through 12th grade. Averaging estimates across ages yields an overall $r = .62$, which is a very large effect size using Cohen's 1988 criteria for correlation coefficients (i.e., $r = .50$). Therefore, given the high redundancy between reading and mathematics, we would expect the NSS to show high, positive correlations with both mathematics and reading scores from other achievement tests—and we further anticipate that the NSS will show somewhat higher correlations with mathematics scores (convergent associations) than with reading scores (divergent associations).

Participants

Participants were drawn from six schools in the same public school district in northern Delaware. The children were part of a longitudinal investigation that evaluated number competencies over six time points, from the beginning of kindergarten to the middle of third grade (Jordan et al., 2009). The schools were selected because they served children from both low-income and middle-income families.

The investigation examined relationships between NSS scores obtained in the fall of kindergarten and reading and mathematics criterion obtained at the end of first grade ($N = 288$) and again at the end of third grade ($N = 211$). In the kindergarten sample, 54% of the children were boys, 56% had minority ethnic backgrounds, and 33% came from low-income families.

Mathematics Criteria

Children were given the Calculation and Applied Problems portions of the Woodcock-Johnson III (McGrew et al., 2007) for a composite mathematics achievement score (WJMath). The Calculation (WJCalc) subtest measures the ability to perform computations in a conventional written format. Applied Problems (WJApp) requires the child to analyze and solve orally presented mathematics problems in various contexts.

Reading Criteria

The Dynamic Indicators of Basic Early Literacy Skills–Sixth Edition (DIBELS; Good & Kaminski, 2002) included measures of letter naming fluency, phoneme segmentation fluency, and nonsense word fluency. The raw score for each measure was the number of letters, phonemes, and nonsense words identified in 1 minute. Scores from the three related measures were totaled for each child and used for the analysis. Average test–retest reliability for the end of kindergarten was .91 (Good, Simmons, Kame'enui, Kaminski, & Wallin, 2002).

The Test of Word Reading Efficiency (Torgesen, Wagner, & Rashotte, 1999) is a standardized measure. In the sight word subtest (Torgesen et al., 1999), students have 45 seconds to read words. The score is the number of correct words. Test–retest and alternate form reliability is $> .90$.

Table 4.12. Convergent and divergent associations for the Number Sense Screener and first-grade criteria

Criterion	Number Sense Screener
Convergent associations	
WJ grade-based calculation standard score	.60[a]
WJ grade-based applied problem standard score	.69
Average[b]	.65
Divergent associations	
DIBELS phoneme segmentation fluency	.12
DIBELS nonsense word fluency	.38
DIBELS oral reading fluency	.51
DIBELS word use fluency	.36
DIBELS retell	.36
Average[b]	.35

Reprinted from *Learning and Individual Differences*, 20, Jordan, N.C., Glutting, J., & Ramineni, C., The importance of number sense to mathematics achievement in first and third grades, 82–88, (2010) with permission from Elsevier.

Note: N = 288; WJ, Woodcock-Johnson; DIBELS, Dynamic Indicators of Basic Early Literacy Skills.

[a]All numbers rounded at second decimal point for convenient presentation.

[b]Average coefficients were calculated with Fisher's *z'* transformation.

Results

Outcomes from analyses completed at the end of first grade are shown in Table 4.12. The NSS showed high correlations (convergent validity) with mathematics scales from the WJMath designed to measure similar attributes. For example, the two correlations between the NSS and WJMath produced an average *r* = .65. The associations were both appreciable and theoretically congruent. Moreover, no scale from the DIBELS reading measure showed a correlation as high. The average correlation between the NSS and the DIBELS was .35. This average also was theoretically congruent, because it was lower than correlations between the NSS and the WJMath. Consequently, it is reasonable to infer from the pattern of convergent and divergent association in Table 4.13 that the NSS shows substantial construct validity.

Table 4.13 shows outcomes obtained at the end of third grade. The pattern was nearly identical to that found in first grade. Once again, a pattern of higher convergent validity (average *r* = .62) was obtained than divergent validity (*r* = .37), with results serving to further enhance assertions that the NSS possesses high levels of construct validity.

Table 4.13. Convergent and divergent associations for the Number Sense Screener and third-grade criteria

Criterion	Number Sense Screener
Convergent associations	
WJ grade-based calculation standard score	.57[a]
WJ grade-based applied problem standard score	.66
Average[b]	.62
Divergent associations	
TOWRE grade-based standard score	.37
Average[b]	.37

Reprinted from *Learning and Individual Differences*, 20, Jordan, N.C., Glutting, J., & Ramineni, C., The importance of number sense to mathematics achievement in first and third grades, 82–88, (2010) with permission from Elsevier.

Note: N = 288; WJ, Woodcock-Johnson; TOWRE, Test of Word Reading Efficiency.

[a]All numbers rounded at second decimal point for convenient presentation.

[b]Average coefficients were calculated with Fisher's *z'* transformation.

Treatment Validity

Dyson, Jordan, and Glutting (in press) examined the effects of an 8-week number sense intervention to develop number competencies among low-income kindergartners ($N = 121$). The intervention purposefully targeted whole-number concepts related to counting, comparing, and manipulating sets. Children were randomly assigned either to a number sense intervention or a "business as usual" contrast group. The intervention was carried out in small-group, 30-minute sessions, 3 days per week for a total of 24 sessions.

The intervention was based on the premise that weaknesses in key number competencies underlie mathematics difficulties and that these competencies can be developed early through purposeful instruction. It targeted number concepts related to counting, comparing, and manipulating sets. The study used a pretest, posttest, and delayed posttest design. Children were randomly assigned either to the intervention condition or a business as usual control group. Dependent variables included a validated assessment of numeracy indicators as well as a standardized measure of mathematics achievement.

Participants

Children were recruited from kindergarten classes in five schools serving high-risk children from low-income urban families. All schools were in the same district. A total of 121 participants completed the study. Fifty-two of the children were girls (43%) and sixty-nine were boys (57%). Sixty-seven of the students were identified as African American (55%), forty-five as Hispanic (37%), seven as Caucasian (6%), one as Asian, and one as biracial, all by teacher report. Thirty of the students (25%) were identified as English language learners and were enrolled in designated kindergarten classrooms for English language learners.

In each of the five participating schools, about half of the participants within each classroom were randomly selected for the intervention, whereas the other half were assigned to a business as usual control group. Because the interventions were carried out in groups of four, there were a few extra children in the various schools who were assigned to the control group, accounting for the unequal numbers in the intervention and control conditions. Participants also were stratified according to kindergarten class.

Measures

The NSS was used as an outcome measure. Mathematics achievement was assessed using the Woodcock-Johnson III Tests of Achievement Form C Brief Battery: Applied Problems and Calculation subtests (WJ; Woodcock, McGrew, Schrank, & Mather, 2007).

Design and Procedures

A pretest, immediate posttest, and delayed posttest design was used. The intervention started in January of 2010 and was carried out in small groups of four children per instructor. The intervention groups met for three 30-minute sessions per week over an 8-week period for a total of 24 lessons.

During the week following the last lesson, children were individually posttested with the NSS and the WJ measures. Approximately 6 weeks later, children were tested again with the same measures.

Intervention

The intervention was designed to augment the regular kindergarten mathematics program with small-group instruction. Skills were reviewed incrementally over the course of the 24 lessons. A compare-and-contrast approach was used throughout the activities. For example, opposites such as before and after, addition and subtraction, $n + 1$ and $n - 1$ were presented simultaneously. The intervention content included the following areas:

1. **Magic number.** To emphasize the structure of the base-10 numeration system, the instructor constructed the magic number using the 10 decade card and single-digit overlays. For example, when the magic number was 11, the instructor placed the 1 card over the 0 in 10 and said, "See, 11 is ten and one more." Eventually, the children picked out the digit card to put over the 0 to make the new magic number. Then, the children counted in unison up to that number. When the number 20 was reached, attention was drawn to the tens place and 20 was shown to be two tens. Numbers after 20 were presented in the same manner as the teens, using the 20 decade card. When 30 was reached, the children were taught the remaining decades up to 90 and were directed to make two-digit numbers using the decade and single digit cards. Because of the difficult naming structure of the numbers 11–19, these numbers were reviewed daily.

2. **Number recognition game.** The magic number was added to the pile of number recognition cards for the daily number recognition game. Whoever got the magic number on his or her turn and named it correctly received a special sticker.

3. **Number sequencing.** Students worked together to put number cards into the proper sequence. Number cards were placed in rows and columns mimicking the 100's chart to highlight that 11 is 10 and 1, 21 is 20 and 1, and so on.

4. **Verbal subitizing.** Children were taught to recognize and name quantities up to 4 instantly, without counting.

5. **Finger use.** Children were taught to count on their fingers and to also make finger quantities to 5 quickly (without counting).

6. **Associating numeral to quantity.** Activities using a number list and chips supported the learning of counting principles related to one-to-correspondence and cardinality.

7. **Number plus (or minus) one principle.** Chips were used with the number list to illustrate that the next number in the count sequence is always one more than the previous number. The $n + 1$ principle leads into the introduction of before and after numbers as $n - 1$ and $n + 1$, respectively. In the field trial, the numbers after were presented first, and then numbers before. Once students learned the after numbers, it was hard to then teach them the before numbers.

8. **Number comparisons.** A cardinality chart was used to support children's understanding of bigger and smaller numbers. The number with the most dots is the bigger of two numbers. The top row of the cardinality chart looks like the number list to reinforce the connection between the count sequence and quantity. Children practiced identifying bigger and smaller numbers through number list and cardinality chart activities.

9. **Part–whole relationships.** Number partners are a pair of numbers whose sum is the given number. Activities using a row of dots and a pencil separator allowed students to discover partners for numbers 2 through 5. Cards were made for each set of partners. By rotating the partner card 180 degrees, the instructor was able to demonstrate the commutative property of addition. Partner card activities included connecting partners to written combinations with Arabic numbers, both for addition and subtraction. Children were shown subtraction problems in the part–whole relationship by having the instructor cover up one part of the partner dot card. This conceptual representation supports children's use of counting up from the subtrahend to solve subtraction problems, a method found easier and more accurate than counting down from the minuend.

10. **Using counting to solve problems.** Children were given opportunities to solve story problems with several strategies, including fingers and writing down tally marks or circles. Children were explicitly shown how to count—on from an addend (counting—on), although they were not discouraged from counting all of their representations in both sets (counting—all) if counting—on was difficult for them. Children were also encouraged to

connect quantities to real-world problems using a magnetic story board and magnetic objects, such as apples or farm animals.

11. **Number board game.** At the end of each session, children played the Great Race Game, adapted from Ramani and Siegler (2008). The game served as a reward for the children working hard during the session. The game board was a colorful number list with numbers 1 to 10. Each number was enclosed in a square, and a starting place was marked at just before the number 1. To determine the number of spaces to move, the child used a spinner divided into two regions, one region containing the number 1 and the other region containing the number 2. After the child spun a 1 or a 2, they were required to "say" their move using a prescribed format that encouraged counting—on.

Results

A series of one-way analyses of covariance (ANCOVA) were conducted to test whether the mean gains between pretest and immediate posttest and delayed posttest for total and subarea scores from the NSS and WJ Achievement Test differed between groups (intervention group versus controls). In addition to p values, effect sizes were reported using standardized beta coefficients (β^S), where values above .05 are considered small but meaningful; those above .10 are considered moderate; and those above .25 are considered large (Keith, 2006).

Although children were randomly assigned to the intervention and control groups, pretest scores from the NSS and/or pretest scores from WJ served as the covariate(s). The covariate(s) served the dual purposes of 1) minimizing any potential confounding that might be attributable to prior mathematics knowledge between the two groups and 2) reducing unexplained variance, and thereby increasing the power of the analyses to detect treatment effects (Field, 2009; Maxwell & Delaney, 2004). Three combinations of covariates were employed: 1) pretest scores from the NSS, 2) pretest scores from WJ, and 3) pretest scores from both the NSS and WJ.

Results were consistent across the ANCOVA models for the NSS and associated subareas; in all instances, the intervention group obtained significantly higher and meaningful adjusted outcome scores at posttest as well as at delayed posttest. For the WJ, the results were significant at posttest but not at delayed posttest, and only for the calculation problems subtest. There were not significant group differences for the applied problems subtest.

These intervention results serve to support and enhance the validity of the NSS by demonstrating that it is sensitive to effective interventions. Specifically, they demonstrate that key areas of number sense can be boosted in kindergartners with established risk for mathematics difficulties or disabilities, many of whom come to school with far fewer learning experiences than their middle-income counterparts. These gains were successfully captured on a number sense assessment tool that is sensitive to short-term progress in kindergarten, and there was some transfer to more conventional written calculation tasks. Although most children seemed to gain from their regular mathematics curriculum, the intervention provided them with added benefits in a relatively short time period.

References

Allen, M.J., & Yen, W.M. (1979). *Introduction to measurement theory*. Monterey, CA: Brooks/Cole.

American Educational Research Association, American Psychological Association, & National Council on Measurement in Education. (1999). *The standards for educational and psychological testing*. Washington, DC: American Educational Research Association.

Anastasi, A., & Urbina, S. (1997). *Psychological testing*. Upper Saddle River, NJ: Prentice Hall.

Baroody, A.J. (1987). The development of counting strategies for single-digit addition. *Journal for Research in Mathematics Education, 18*(2), 141–157.

Baroody, A.J., Eiland, M., & Thompson, B. (2009). Fostering at-risk preschoolers' number sense. *Early Education & Development, 20*(1), 49.

Baroody, A.J., Ginsburg, H.P., & Waxman, B. (1983). Children's use of mathematical structure. *Journal for Research in Mathematics Education, 14*(3), 156–168.

Baroody, A.J., Lai, M.-L., & Mix, K.S. (2006). The development of young children's early number and operation sense and its implications for early childhood education. In B. Spodek & O. Saracho (Eds.), *Handbook of research on the education of young children* (pp. 187–221). Mahwah, NJ: Lawrence Erlbaum Associates.

Baroody, A.J., & Rosu, L. (2006). *Adaptive expertise with basic addition and subtraction combinations: The number sense view*. Paper presented at the meeting of the American Educational Research Association, San Francisco.

Bond, T.G., & Fox, C.M. (2007). *Applying the Rasch model: Fundamental measurement in the human sciences*. Mahwah, NJ: Lawrence Erlbaum Associates.

Booth, J.L., & Siegler, R.S. (2006). Developmental and individual differences in pure numerical estimation. *Developmental Psychology, 41*, 189–201.

Butterworth, B., & Reigosa, V. (2007). Information processing deficits in dyscalculia. In D.B. Berch & M.M.M. Mazzocco (Eds.), *Why is math so hard for some children? The nature and origins of mathematical learning difficulties and disabilities* (pp. 107–120). Baltimore: Paul H. Brookes Publishing Co.

Campbell, D.L. (1960). Recommendations for APA test standards regarding construct, trait, and discriminant validity. *American Psychologist, 15*, 546–553.

Campell, D.T., & Fiske, D.W. (1959). Convergent and discriminant validation by the multitrait-multimethod matrix. *Psychological Bulletin, 56*, 81–105.

Carmines, E.G., & Zeller, R.A. (1979). *Reliability and validity assessment*. Newbury Park, CA: Sage.

Case, R., & Griffin, S. (1990). Child cognitive development: The role of central conceptual structures in the development of scientific and social thought. In E.A. Hauert (Ed.), *Developmental psychology: Cognitive, perceptuo-motor, and neurological perspectives* (pp. 193–230). Amsterdam, The Netherlands: Elsevier.

Cohen, J. (1988). *Statistical power analysis for the behavioral sciences*. Hillsdale, NJ: Lawrence Erlbaum Associates.

Cronbach, L.J. (1951). Coefficient alpha and the internal structure of tests. *Psychometrika, 16*, 297–334.

Cronbach, L.J., & Meehl, P.H. (1955). Construct validity in psychological tests. *Psychological Bulletin, 52*, 281–302.

Delaware Department of Education. (2008). *Delaware state testing program technical report—2007*. Dover, DE: Author.

Dorans, N.J., & Holland, P.W. (1993). DIF detection and description: Mantel-Haenszel and standardization. In P.W. Holland & H. Wainer (Eds.), *Differential item functioning* (pp. 35–66). Hillsdale, NJ: Lawrence Erlbaum.

Dudek, F.J. (1979). The continuing misinterpretation of the standard error of measurement. *Psychological Bulletin, 86*, 335–337.

Duncan, G.J., Dowsett, C.J., Claessens, A., Magnuson, K., Huston, A.C., Klebanov, P., et al. (2007). School readiness and later achievement. *Developmental Psychology, 43*(6), 1428–1446.

Dyson, N.I., Jordan, N.C., & Glutting, J. (in press). A number sense intervention for urban kindergarteners at-risk for mathematics difficulties. *Journal of Learning Disabilities*.

Embretson, S.E., & Reise, S.P. (2000). *Item response theory for psychologists.* Mahwah, NJ: Lawrence Erlbaum.

Field, A. (2009). *Discovering statistics using SPSS* (3rd ed.). London: Sage.

Fuson, K.C., Grandau, L., & Sugiyama, P.A. (2001). Achievable numerical understandings for all young children. *Teaching Children Mathematics, 7*(9), 522–526.

Geary, D.C. (1995). Reflections of evolution and culture in children's cognition: Implications for mathematical development and instruction. *American Psychologist, 50,* 24–37.

Geary, D.C., Hamson, C.O., & Hoard, M.K. (2000). Numerical and arithmetical cognition: A longitudinal study of process and concept deficits in children with learning disability. *Journal of Experimental Child Psychology, 77*(3), 236–263.

Geary, D.C., Hoard, M.K., Byrd-Craven, J., Nugent, L., & Numtee, C. (2007). Cognitive mechanisms underlying achievement deficits in children with mathematical learning disability. *Child Development, 78*(4), 1343.

Gelman, R., & Butterworth, B. (2005). Number and language: How are they related? *Trends in Cognitive Sciences, 9,* 6–10.

Gelman, R., & Gallistel, C.R. (1978). *The child's understanding of number.* Cambridge, MA: Harvard University Press.

Gersten, R., Jordan, N.C., & Flojo, J.R. (2005). Early identification and interventions for students with mathematics difficulties. *Journal of Learning Disabilities, 38*(4), 293–304.

Ginsburg, H.P. (1989). *Children's arithmetic.* Austin, TX: PRO-ED.

Ginsburg, H.P., & Golbeck, S.L. (2004). Thoughts on the future of research on mathematics and science learning and education. *Early Childhood Research Quarterly, 19(1),*190–200.

Ginsburg, H.P., & Russell, R.L. (1981). Social class and racial influences on early mathematical thinking. *Monographs of the Society for Research in Child Development, 46*(6, Serial No. 69).

Glutting, J.J., McDermott, P.A., & Stanley, J.C. (1987). Resolving differences among methods of establishing confidence limits for test scores. *Educational and Psychological Measurement, 47,* 607–614.

Good, R.H., & Kaminski, R.A. (Eds.). (2002). *Dynamic Indicators of Basic Early Literacy Skills (DIBELS).* Eugene, OR: Institute for the Development of Educational Achievement.

Good, R.H., Simmons, D., Kame'enui, E., Kaminski, R.A., & Wallin, J. (2002). *Summary of decision rules for intensive, strategic, and benchmark instructional recommendations in kindergarten through third grade* (Technical Report No. 11). Eugene, OR: University of Oregon.

Gregory, R.J. (2007). *Psychological testing: History, principles, and applications.* Boston: Allyn & Bacon.

Griffin, S. (2002). The development of math competence in the preschool and early school years: Cognitive foundations and instructional strategies. In J.M. Roher (Ed.), *Mathematical cognition* (pp. 1–32). Greenwich, CT: Information Age Publishing.

Griffin, S. (2004). Building number sense with Number Worlds: A mathematics program for young children. *Early Childhood Research Quarterly, 19*(1), 173–180.

Griffin, S., & Case, R. (1997). Rethinking the primary school math curriculum: An approach based on cognitive science. *Issues in Education, 3*(1), 1–49.

Griffin, S., Case, R., & Siegler, R. (1994). Rightstart: Providing the central conceptual prerequisites for first formal learning of arithmetic to students at risk for school failure. In K. McGilly (Ed.), *Classroom lessons: Integrating cognitive theory and classroom practice* (pp. 24–49). Cambridge, MA: MIT Press.

Huttenlocher, J., Jordan, N.C., & Levine, S.C. (1994). A mental model for early arithmetic. *Journal of Experimental Psychology, 123,* 284–296.

Individuals with Disabilities Education Improvement Act (IDEA) of 2004, PL 108-446, 20 U.S.C. §§ 1400 *et seq.*

Jordan, N.C. (2007). The need for number sense. *Educational Leadership, 65*(2), 63–66.

Jordan, N.C., Glutting, J., & Ramineni, C. (2008). A number sense assessment tool for identifying children at risk for mathematical difficulties. In A. Dowker (Ed.), *Mathematical difficulties: Psychology and intervention* (pp. 45–58). San Diego: Academic Press.

Jordan, N.C., Glutting, J., & Ramineni, C. (2009). The importance of number sense to mathematics achievement in first and third grades. *Learning and Individual Differences, 20*(2), 82–88.

Jordan, N.C., Glutting, J., Ramineni, C., & Watkins, M.W. (2010). Validating a number sense screening tool for use in kindergarten and first grade: Prediction of mathematics proficiency in third grade. *School Psychology Review, 39(2),* 181–195.

Jordan, N.C., Hanich, L.B., & Kaplan, D. (2003a). A longitudinal study of mathematical competencies in children with specific mathematics difficulties versus children with co-morbid mathematics and reading difficulties. *Child Development, 74*(3), 834–850.

Jordan, N.C., Hanich, L.B., & Kaplan, D. (2003b). Arithmetic fact mastery in young children: A longitudinal investigation. *Journal of Experimental Child Psychology, 85,* 103–119.

Jordan, N.C., Huttenlocher, J., & Levine, S.C. (1992). Differential calculation abilities in young children from middle- and low-income families. *Developmental Psychology, 28*(4), 644–653.

Jordan, N.C., Kaplan, D., Locuniak, M.N., & Ramani, G.B. (2007). Predicting first-grade math achievement from developmental number sense trajectories. *Learning Disabilities Research & Practice, 22*(1), 36–46.

Jordan, N.C., Kaplan, D., Olah, L., & Locuniak, M.N. (2006). Number sense growth in kindergarten: A longitudinal investigation of children at risk for mathematics difficulties. *Child Development, 77*, 153–175.

Jordan, N.C., Kaplan, D., Ramineni, C., & Locuniak, M.N. (2008). Development of number combination skill in the early school years: When do fingers help? *Developmental Science,11*(5), 662–668.

Jordan, N.C., Kaplan, D., Ramineni, C., & Locuniak, M.N. (2009). Early math matters: Kindergarten number competence and later mathematics outcomes. *Developmental Psychology, 45*(3), 850–867.

Jordan, N.C., & Levine, S.C. (2009). Socioeconomic variation, number competence, and mathematics learning difficulties in young children. *Developmental Disabilities Research Reviews, 15*(1), 60–68.

Jordan, N.C., Levine, S.C., & Huttenlocher, J. (1993). Differential calculation abilities in young children at risk: Linking research with assessment and intervention. In N.C. Jordan & J. Goldsmith-Phillips (Eds.), *Learning disabilities: New directions for assessment and intervention*. Boston: Allyn & Bacon.

Jordan, N.C., Levine, S.C., & Huttenlocher, J. (1994). Development of calculation abilities in middle- and low-income children after formal instruction in school. *Journal of Applied Developmental Psychology, 15*(2), 223–240.

Keith, T.Z. (2006). *Multiple regression and beyond*. Boston: Pearson.

Keppel, G., & Wickens, T.D. (2004). *Design and analysis: A researcher's handbook* (4th ed.). Englewood Cliffs, NJ: Prentice Hall.

Landerl, K., Bevan, A., & Butterworth, B. (2004). Developmental dyscalculia and basic numerical capacities: A study of 8–9-year-old students. *Cognition, 93*(2), 99–125.

Levine, S.C., Jordan, N.C., & Huttenlocher, J. (1992). Development of calculation abilities in young children. *Journal of Experimental Child Psychology, 53*, 72–103.

Linacre, J.M. (1998). Rasch first or factor first? *Rasch Measurement Transactions, 11*, 603.

Linacre, J.M. (2007). *A user's guide to FACETS: Rasch measurement computer program*. Chicago: MESA Press.

Locuniak, M.N., & Jordan, N.C. (2008). Using kindergarten number sense to predict calculation fluency in second grade. *Journal of Learning Disabilities, 41*(5), 451–459.

Malofeeva, E., Day, J., Saco, X., Young, L., & Ciancio, D. (2004). Construction and evaluation of a number sense test with Head Start children. *Journal of Educational Psychology, 96*(4), 648–659.

Mantel, N., & Haenszel, W. (1959). Statistical aspects of the analysis of data from retrospective studies of disease. *Journal of the National Cancer Institute, 22*, 719–748.

Maxwell, S.E., & Delaney, H.D. (2004). *Designing experiments and analyzing data: A model comparison perspective* (2nd ed.). Mahwah, NJ: Lawrence Erlbaum Associates.

McGrew K.S., Schrank F.A., & Woodcock R.W. (2007). *Woodcock-Johnson III normative update*. Rolling Meadows, IL: Riverside Publishing.

Messick, S. (1989). Validity. In R.L. Linn (Ed.), *Educational measurement* (3rd ed., pp. 13–103). New York: Macmillan.

Mix, K.S., Huttenlocher, J., & Levine, S.C. (2002). *Quantitative development in infancy and early childhood*. New York: Oxford University Press.

Murphy, K.R., & Myors, B. (2004). *Statistical power analysis: A simple and general model for traditional and modern hypothesis tests*. Mahwah, NJ: Lawrence Erlbaum Associates.

National Center for Education Statistics. (2009). *The nation's report card: Mathematics 2009* (No. NCES 2010-451). Washington, DC: Author.

National Council of Teachers of Mathematics. (2006). *Curriculum focal points for prekindergarten through grade 8 mathematics: A quest for coherences*. Reston, VA: Author.

National Governors Association Center for Best Practices & Council of Chief State School Officers. (2010). *Common core state standards for mathematics*. Retrieved November 15, 2010, from http://www.corestandards.org/assets/CCSSI_Math%20Standards.pdf

National Mathematics Advisory Panel. (2008). *Foundations for success: The final report of the National Mathematics Advisory Panel*. Washington, DC: U.S. Department of Education.

National Research Council. (2009). *Mathematics learning in early childhood: Paths toward excellence and equity*. Washington, DC: National Academies Press.

Ramani, G.B., & Siegler, R.S. (2008). Promoting broad and stable improvements in low-income children's numerical knowledge through playing number board games. *Child Development, 79*(2), 375–394.

Rasch, G. (1960). *Probabalistic models for some intelligence and attainment tests*. Chicago: University of Chicago.

Reynolds, C.R., Livingston, R.B., & Willson, V. (2006). *Measurement and assessment in education*. Boston: Allyn & Bacon.

Sadler, P.M., & Tai, R.H. (2007). Weighting for recognition: Accounting for advanced placement and honors courses when calculating high school grade point average. *National Association of Secondary School Principals Bulletin, 91*, 5–32.

Salvia, J., Ysseldyke, J.E., & Bolt, S. (2009). *Assessment in special and inclusive education*. Belmont, CA: Cengage Learning.

Siegler, R.S., & Booth, J.L. (2004). Development of numerical estimation in young children. *Child Development, 75*, 428–444.

Siegler, R.S., & Shrager, J. (1984). Strategy choices in addition and subtraction: How do children know what to do? In C. Sophian (Ed.), *The origins of cognitive skills* (pp. 229–293). Hillsdale, NJ: Erlbaum.

Starkey, P., & Cooper, R. (1980). Perception of numbers by human infants. *Science, 210*(28), 1033–1034.

Starkey, P., Klein, A., & Wakeley, A. (2004). Enhancing young children's mathematical knowledge through a pre-kindergarten mathematics intervention. *Early Childhood Research Quarterly, 19*(1), 99–120.

Thomas, M.S.C., & Karmiloff-Smith, A. (2003). Modeling language acquisition in atypical phenotypes. *Psychological Review, 110*(4), 647–682.

Thorndike, R.L. (1982). *Applied psychometrics*. Boston: Houghton Mifflin.

Torgesen, J.K., Wagner, R.K., & Rashotte, C.A. (1999). *TOWRE, Test of Word Reading Efficiency*. Austin, TX: PRO-ED.

VanDerHeyden, A.M., Broussard, C., & Cooley, A. (2006). Further development of measures of early math performance for preschoolers. *Journal of School Psychology, 44*(6), 533–553.

Wechsler, D. (1999). *Wechsler Intelligence Scale for Children* (3rd ed.). San Antonio, TX: Psychological Corporation.

Wechsler, D. (2009). *Wechsler Individual Achievement Test* (2nd ed.). San Antonio, TX: Psychological Corporation.

Woodcock, R.W., McGrew, K.S., Schrank, F.A., & Mather, N. (2007). *Woodcock-Johnson III normative update*. Rolling Meadows, IL: Riverside Publishing.

Wright, B.D., & Linacre, J.M. (1994). Reasonable mean-square fit values. *Rasch Measurement Transactions, 8*(3), 370.

Wright, B.D., & Masters, G.N. (1982). *Rating scale analysis*. Chicago: MESA Press.

Wright, B.D., & Stone, M. (1979). *Best test design*. Chicago: MESA Press.

Wynn, K. (1992). Children's acquisition of the number words and the counting system. *Cognitive Psychology, 24*, 220–251.

Instructions for Making the Nonverbal Calculation Materials

White Mat

The mat is a piece of white craft foam (approximately 9 × 12 inches) that can be purchased at most craft stores.

Box

Materials

- A box and its cover, similar to materials used for storing personal checks
- Solid-colored contact paper

Instructions

Cut a small triangle out of the short side of a box cover (see diagram) leaving approximately 1½ inches open at the bottom of the triangle. The opening must be large enough to allow the dots (tokens) to slide into the box.

If necessary, wrap the box cover with solid-colored (not neon) contact paper. No words or pictures should be visible on the box cover. Keep extra dots in the box bottom; use only the box cover for the Nonverbal Calculations. The box and cover make a handy carrying case for the dots (tokens).

Dots

Black dots (tokens) can be purchased from web sites that sell game pieces. The dots (tokens) should be plain and not decorated in any way that would be distracting to the student.

Story Problems and Number Combinations Worksheet

Story Problems and Number Combinations Worksheet

Name _____

School _____   Date _____

Story Problems

Number Combinations

Number List
Worksheet

| 1 | 2 | 3 | 4 | 5 | 6 | 7 | 8 | 9 | 10 |

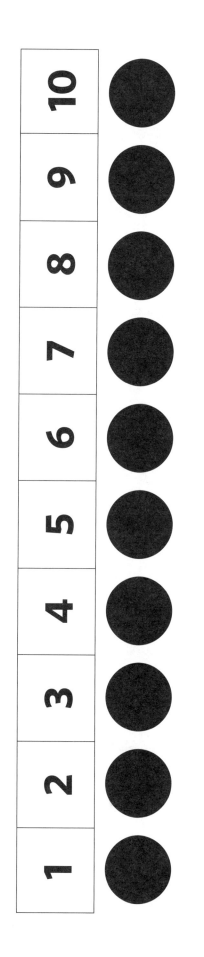

- - - - Cut here - - - -

| 1 | 2 | 3 | 4 | 5 | 6 | 7 | 8 | 9 | 10 |

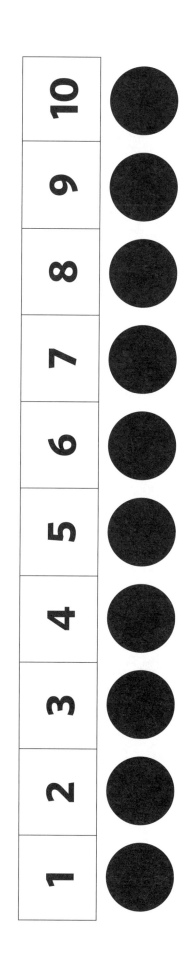

Converting Raw Scores to Standard Scores and Percentile Ranks

Table 1. Norms for fall of kindergarten

Raw score/sum of raw scores	Standard score (*M* = 100, *SD* = 15)	Percentile
0	50	0.04
1	50	0.04
2	51	0.05
3	51	0.05
4	51	0.05
5	52	0.07
6	65	1
7	72	3
8	76	5
9	81	10
10	86	18
11	89	23
12	93	32
13	97	42
14	99	47

(continued)

Table 1. *(continued)*

15	101	53
16	105	63
17	107	68
18	111	77
19	113	81
20	114	82
21	116	86
22	120	91
23	124	95
24	131	98
25	138	99
≥ 26	145	99.87

Table 2. Norms for spring of kindergarten

Raw score/sum of raw scores	Standard score ($M = 100$, $SD = 15$)	Percentile
0	45	0.01
1	45	0.01
2	46	0.02
3	46	0.02
4	46	0.02
5	47	0.02
6	50	0.04
7	68	2
8	75	5
9	80	9

(continued)

Table 2.   *(continued)*

10	82	12
11	85	16
12	88	21
13	92	30
14	94	34
15	96	39
16	99	47
17	102	55
18	103	58
19	106	66
20	109	73
21	111	77
22	113	81
23	115	84
24	121	92
25	126	96
$\geq$ 26	133	99

Table 3.   Norms for fall of first grade

Raw score/sum of raw scores	Standard score (M = 100, SD = 15)	Percentile
0	45	0.01
1	46	0.02
2	47	0.02
3	47	0.02
4	48	0.03

(continued)

Table 3. (*continued*)

5	49	0.03
6	50	0.04
7	51	0.05
8	52	0.07
9	52	0.07
10	65	1
11	67	1
12	71	3
13	76	5
14	80	9
15	82	12
16	84	14
17	85	16
18	86	18
19	88	21
20	90	25
21	92	30
22	94	34
23	96	39
24	97	42
25	100	50
26	105	63
27	110	75
28	118	88
29	128	97

Index

Tables and figures are indicated by *t* and *f*, respectively.